ok

Cloning

ISSUES

(formerly Issues for the Nineties)

Volume 12

Editor

Craig Donnellan

Independence

Educational Publishers

Cambridge

First published by Independence
PO Box 295
Cambridge CB1 3XP
England

© Craig Donnellan 2000

British Library Cataloguing in Publication Data
Cloning – (Issues Series)
I. Donnellan, Craig II. Series
174.9'57

ISBN 1 86168 116 X

Printed in Great Britain
The Burlington Press
Cambridge

Typeset by
Claire Boyd

Cover
The illustration on the front cover is by
Pumpkin House.

CONTENTS

Chapter One: An Overview

Chapter Two: The Cloning Debate

Introduction

Cloning is the twelfth volume in the **Issues** series. The aim of this series is to offer up-to-date information about important issues in our world.

Cloning looks at the current debate, the benefits and concerns about cloning.

The information comes from a wide variety of sources and includes:
Government reports and statistics
Newspaper reports and features
Magazine articles and surveys
Literature from lobby groups
and charitable organisations.

It is hoped that, as you read about the many aspects of the issues explored in this book, you will critically evaluate the information presented. It is important that you decide whether you are being presented with facts or opinions. Does the writer give a biased or an unbiased report? If an opinion is being expressed, do you agree with the writer?

Cloning offers a useful starting-point for those who need convenient access to information about the many issues involved. However, it is only a starting-point. At the back of the book is a list of organisations which you may want to contact for further information.

*****

Cloning

Information from *New Scientist*

Is cloning 'unnatural'?

Not at all – some organisms in nature only reproduce using cloning – not only bacteria and yeasts, but also larger organisms like some snails and shrimp. Because in nature sexual reproduction is the only way to improve the genetic stock of a species, most asexual species tend to die off, but at least one – a shrimp called *Artemia perthenogenetica* – has survived for at least 30 million years. Many more species, including the aphid, reproduce by cloning most of the time, only reproducing sexually every few generations. Perhaps one day humankind may follow their lead.

Is an identical twin essentially the same as a clone?

Only if the clone is born at the same time from the same womb as its clone, as we now know that what a foetus is exposed to in the womb, in the way of nutrition or alcohol or drugs or perhaps even stress hormones, can influence its physical and mental development.

Could some lunatic clone Hitler if human cloning were perfected?

Just possibly – but they wouldn't get what they wanted. First, they would need some living cells from his body – unless it was frozen or otherwise preserved soon after death they would probably be unusable. More importantly, because of differences in the environment of the womb and upbringing clone Hitlers would not act, think or even necessarily look like the original.

Could clones be 'farmed' to provide spare body parts for their 'parent' clone without problems of tissue rejection?

Possibly, although we don't know enough yet to be confident that rejection would be eliminated entirely. You would also have to wait a number of years until the clone's organs were mature enough to transplant, and of course your actions would be highly illegal unless your clone was willing to act as a donor as a clone would be just as human as you or I. Even leaving aside the ethical concerns, with the progress that is being made in understanding and coping with tissue rejection, you would be more likely to have a pig's heart in your future than a clone's.

Would a clone have a soul?

Though we are not theologians if you grant souls to identical twins and to the various kinds of 'test-tube babies' already being born then it follows that a clone would have one too.

Could people be cloned without conscious brains (so their body parts could be harvested with fewer moral qualms)?

No. For starters, whatever consciousness is, it doesn't reside in any one brain structure or set of genes that could be easily removed from the clone before or during its development. Moreover, attempting to surgically or genetically erase someone's 'consciousness' is itself morally dubious. It would also be hard to know if your 'technique' worked. A person can look and behave like a mindless vegetable but have a very active mind – witness the paralysed French writer, Jean-Dominique Bauby, who dictated a 130-page novel by moving an eyelid.

Could vital organs be grown using cloning without the rest of a body?

Possibly – but nobody is even close to knowing how. Contrary to scientists' expectations, the birth of Dolly shows it is possible to reprogramme the cell of an adult (or at least its genome) so that it begins development all over again. This newly discovered flexibility means it may one day be possible to reprogramme skin or blood cells so that they grow into 'spare part' tissues

CLONE RESEARCH DEPT.

and organs, rather than whole organisms. But the technical obstacles will be huge.

Could cloning be used to create 'super warriors' or super-intelligent people?

Possibly – though we don't yet know enough about human genetics to do much 'improving' of people. So far, because of ethical concerns, geneticists are concentrating on finding the causes of genetic diseases and then curing them. While cloning makes it easier to meddle with human and animal genes, even before recent discoveries a considerable amount of genetic improvement of animals was already taking place. A thoroughbred horse is essentially genetically engineered, for example.

Could cloning be used to save endangered species?

At the moment its success rate is very low (Dolly was only cloned after 276 tries) but if this can be improved on it might well turn out to be useful to increase the population of hard-to-breed animals. Extinct animals (or animals without females) would be more difficult. A female can't normally give birth to an animal of a different species, but it is not yet clear whether a female of a closely-related species could give birth to a clone of a different species.

Could cloning help gay couples to conceive and make men unnecessary for procreation?

In principle, yes. Of course a clone would have to be the identical twin of one or the other partner – it would be difficult to duplicate any of the mixing of genes that occurs during sexual reproduction using cloning techniques.

A history of cloning

5000 BC – A better breed of corn
Early humans discover that if they plant seeds produced by the heartiest plants, the next crop will be a strong one. This is the first step in manipulating life to suit human needs – the ultimate goal of cloning.

1000 BC – Mules, the original hybrid
To ease their backbreaking labour, farmers breed female horses with male asses to produce mules. Mules are desirable beasts of burden because of their endurance, surefootedness, intelligence and long lifespan.

1952 – A tadpole is cloned
A tiny tadpole makes history as the first cloned animal. Using cells from a tadpole embryo, Robert Briggs and Thomas King create new tadpoles identical to the original donor.

1965 – Cloning goes camp
Cloning goes camp in the B-movie classic, *The Human Duplicators*. Primitive special effects dominate this hilarious alien-invaders flick. The plot is somewhat difficult to follow, but the low-budget shots of the alien spacecraft and the cloning process make this a must-see. Can't find it at your local video store? Try *Jaws of the Alien*, it's the same movie.

1972 – Xeroxing a gene
Cloning steps down to the minute level with the first cloning of a gene. Scientists isolate the gene, then bind it to an organism (in this instance a yeast) that incorporates the gene into its own DNA and multiplies, producing many copies of the desired gene.

1976 – From mice to men
Rudolf Jaenisch of the Salk Institute for Biological Studies in La Jolla, California, injects human DNA into newly fertilised mouse eggs to produce mice that are part human. When the mice reproduce, they pass their human genetic material to their offspring, creating a slew of so-called transgenic mice. Different human diseases can be studied by creating mice with the appropriate genetic composition.

1978 – The world's first test-tube baby
The world clamours for a glimpse of Baby Louise, the first child conceived through in-vitro fertilisation. Using the husband's sperm, British doctors fertilise an egg in a petri dish, then implant the embryo in the uterus of the healthy woman.

1978 – A tale of human cloning
David Rorvik's book, *In His Image: The Cloning of a Man*, is published. In the novel, a man is cloned from

skin cells. But this would be scientifically impossible: although skin cells contain all of the genes needed for an embryo to develop, only the skin genes are active. Thus, all that would be reproduced would be more skin cells.

1987 – From embryo to ewe

The first mammals, sheep and cows, are cloned from embryonic cells. But animals cloned from embryonic cells contain the genetic material of both parents because the embryos are sexually fertilised. Clones from embryonic cells from the same parents fertilised at different times are as different as brothers and sisters.

1993 – Replicating T. Rex

Dino-clones appear in *Jurassic Park*. A few strands of prehistoric DNA have fossilised, and that is enough for an ambitious theme-park owner to wreak genetic havoc. Any plot in Steven Spielberg's money-churning franchise is merely an excuse to string together amazing special-effects scenes of raptors running rampant, but that does little to detract from the most successful cloning movie ever made.

1993 – 'All about Eve'

Clones show up on *The X-Files*. Television's best-loved FBI agents are assigned to a mysterious case in which identical girls on opposite coasts each wind up with a dead parent. It turns out the girls share a secret – and that is not all they share, either. A genuinely spooky episode of the cult-favourite show – demand among viewers was so high that Fox has marketed a videotape of the programme.

1995 – Sequencing the genome

Sequencing all of a bacterium's genetic material is no easy task for J. Craig Venter of the Institute for Genomic Research in Rockville, Md. Venter uses chemical analysis to identify all the genes of *Hemophilus influenzae*, the bacterium that causes meningitis and children's ear infections. Scientists hope to sequence the human genome by 2005, which would enable scientists to target specific genes for research.

1996 – Keaton's comedic cloning

Michael Keaton stars in the harmless comedy *Multiplicity*, as a contractor unable to juggle the demands of his busy life. The solution? A mad scientist makes a few extra versions of him to go around. The film skips over any messy scientific detail, but the sight of Keaton playing off multiple versions of himself is enough to give pause to anyone considering lending themselves an extra pair of hands.

1997 – Hello, Dolly

Dolly is cloned and born in 1996, but the world doesn't say hello to Dolly until 1997, when her existence is revealed. Using older techniques as well as some new tricks, Scottish researcher Ian Wilmut clones Dolly from an adult cell. He hopes to use this scientific advancement to clone transgenic livestock.

Views

People and organisations

In the debate over cloning, there are those that feel that the benefits and advances gained from cloning outweigh any social dilemmas, and there are those that feel that cloning is wrong on a fundamental moral level and would produce scientific and social problems. In weighing in on these views, major organisations draw on numerous sources including religious law, party philosophy and scientific concern.

Some object to cloning on a purely ethical level, while others favour cloning solely for the scientific advances it will produce. These are the stances of some prominent religious, scientific, and ethical groups.

The Catholic Church

John Paul II released a statement condemning the cloning of all life forms. The Vatican also issued a statement that only condemned human cloning, but did not address other forms.

Sunni Islam

Abdelmo'ti Bayyumi, theologian from Al-Azhar University, declared it is forbidden to clone animals under Islamic law. However, some Muslims have testified to the National Bioethics Advisory Committee that they feel cloning might be allowable if it produced ways to counteract infertility.

Biotechnology Industry Organisation

Carl Felbaum, president: 'One of the prospects should not be, perhaps should never be, the extension of this technique to human beings . . . Now that it may be possible we would say it should be prohibited if necessary by law.'

Stephen Grebe

Professor of biology, American University: 'We're going to be facing this issue with humans . . . With that possibility open, I'm concerned that without safeguards this will become a reality. It may very well already be.'

Church of England Board of Social Responsibility:

Mary Seller: 'The antics of a few cranks and Hitler types' should not impede cloning research. 'Cloning, like all science, must be used responsibly. Cloning humans is not desirable. But cloning sheep has its uses.'

World Council of Churches

Martin Robra, executive secretary: The council would prefer a moratorium until all ethical questions can be resolved.

Cloning techniques

Information from *Conceiving a Clone* – Kayotic Development

Nuclear transfer (a general overview)

First explored by Hans Spemann in the 1920s to conduct genetics research, nuclear transfer is the technique currently used in the cloning of adult animals. A technique known as twinning exists, but can only be used before an organism's cells differentiate. All cloning experiments of adult mammals have used a variation of nuclear transfer.

Nuclear transfer requires two cells, a donor cell and an oocyte, or egg cell. Research has proven that the egg cell works best if it is unfertilised, because it is more likely to accept the donor nucleus as its own. The egg cell must be enucleated. This eliminates the majority of its genetic information. The donor cell is then forced into the Gap Zero, or G0 cell stage, a dormant phase, in different ways depending on the technique. This dormant phase causes the cell to shut down but not die. In this state, the nucleus is ready to be accepted by the egg cell. The donor cell's nucleus is then placed inside the egg cell, either through cell fusion or transplantation. The egg cell is then prompted to begin forming an embryo. When this happens, the embryo is then transplanted into a surrogate mother. If all is done correctly, occasionally a perfect replica of the donor animal will be born.

Each group of researchers has its own specific technique. The best-known is the Roslin technique, and the most effective and most recently developed is the Honolulu technique.

The Roslin technique

The cloning of Dolly has been the most important event in cloning history. Not only did it spark public interest in the subject, but it also proved that the cloning of adult animals could be accomplished. Previously, it was not known if an adult nucleus was still able to produce a completely new animal. Genetic damage and the simple deactivation of genes in cells were both considered possibly irreversible.

The realisation that this was not the case came after the discovery by

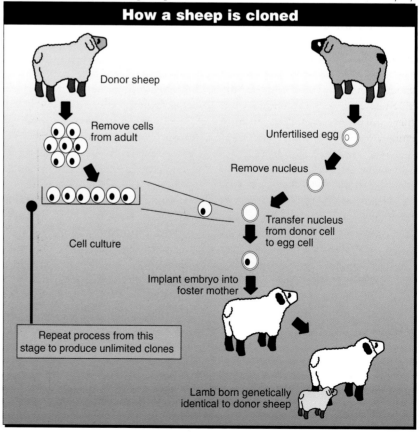

How a sheep is cloned

Donor sheep

Remove cells from adult

Cell culture

Repeat process from this stage to produce unlimited clones

Unfertilised egg

Remove nucleus

Transfer nucleus from donor cell to egg cell

Implant embryo into foster mother

Lamb born genetically identical to donor sheep

Ian Wilmut and Keith Cambell of a method with which to synchronise the cell cycles of the donor cell and the egg cell. Without synchronised cell cycles, the nucleus would not be in the correct state for the embryo to accept it. Somehow the donor cell had to be forced into the Gap Zero, or G0 cell stage, or the dormant cell stage.

First, a cell (the donor cell) was selected from the udder cells of a Finn Dorset sheep to provide the genetic information for the clone. For this experiment, the researchers allowed the cell to divide and form a culture *in vitro*, or outside of an animal. This produced multiple copies of the same nucleus. This step only becomes useful when the DNA is altered, such as in the case of Polly, because then the changes can be studied to make sure that they have taken effect.

A donor cell was taken from the culture and then starved in a mixture which had only enough nutrients to keep the cell alive. This caused the cell to begin shutting down all active genes and enter the G0 stage. The egg cell of a Blackface ewe was then enucleated and placed next to the donor cell. One to eight hours after the removal of the egg cell, an electric pulse was used to fuse the two cells together and, at the same time, activate the development of an embryo. This technique for mimicking the activation provided by sperm is not completely correct, since only a few electrically activated cells survive long enough to produce an embryo.

If the embryo survives, it is allowed to grow for about six days, incubating in a sheep's oviduct. It has been found that cells placed in oviducts early in their development are much more likely to survive than those incubated in the lab. Finally, the embryo is placed into the uterus of a surrogate mother ewe. That ewe then carries the clone until it is ready to give birth. Assuming nothing goes wrong, an exact copy of the donor animal is born.

This newborn sheep has all of the same characteristics of a normal newborn sheep. It has yet to be seen if any adverse effects, such as a higher risk of cancer or other genetic diseases

that occur with the gradual damage to DNA over time, are present in Dolly or other animals cloned with this method.

The Honolulu technique

In July of 1998, a team of scientists at the University of Hawaii announced that they had produced three generations of genetically identical cloned mice.

The technique is accredited to Teruhiko Wakayama and Ryuzo Yanagimachi of the University of Hawaii. Mice had long been held to be one of the most difficult mammals to clone due to the fact that almost immediately after a mouse egg is fertilised, it begins dividing. Sheep were used in the Roslin technique because their eggs wait several hours before dividing, possibly giving the egg time to reprogramme its new nucleus. Even without this luxury, Wakayama and Yanagimachi were able to clone with a much higher success rate (three clones out of every one-hundred attempts) than Ian Wilmut (one in 277).

Wakayama approached the problem of synchronising cell cycles differently from Wilmut. Wilmut used udder cells, which had to be forced into the G0 stage. Wakayama initially used three types of cells, Sertoli cells, brain cells, and cumulus cells. Sertoli and brain cells both

remain in the G0 state naturally and cumulus cells are almost always in either the G0 or G1 state.

Unfertilised mouse egg cells were used as the recipients of the donor nuclei. After being enucleated, the egg cells had donor nuclei inserted into them. The donor nuclei were taken from cells within minutes of each cell's extraction from a mouse. Unlike the process used to create Dolly, no *in vitro*, or outside of an animal, culturing was done on the cells. After one hour, the cells had accepted the new nucleus. After an additional five hours, the egg cell was then placed in a chemical culture to jump-start the cell's growth, just as fertilisation does in nature.

In the culture was a substance (cytochalasin B) which stopped the formation of a polar body, a second cell which normally forms before fertilisation. The polar body would take half of the genes of the cell, preparing the other cell to receive genes from sperm.

After being jump-started, the cells develop into embryos. These embryos can then be transplanted into surrogate mothers and carried to term. The most successful of the cells for the process were cumulus cells, so research was concentrated on cells of that type.

After proving that the technique was viable, Wakayama also made clones of clones and allowed the original clones to give birth normally to prove that they had full reproductive functions. At the time he released his results, Wakayama had created fifty clones.

This new technique allows for further research into exactly how an egg reprogrammes a nucleus, since the cell functions and genomes of mice are some of the best understood. Mice also reproduce within months, much more rapidly than sheep. This aids in researching long-term results.

Cloning timeline

Information from *Conceiving a Clone* – Kayotic Development

Earlier years

1885
August Weissmann states genetic information of a cell diminishes with each cell division.

1902
Walter Sutton proves chromosomes hold genetic information.
Hans Spemann divides a Salamander embryo in two and shows early embryo cells retain all the genetic information necessary to create a new organism.

1928
Hans Spemann performs first nuclear transfer experiment.

1938
Hans Spemann proposes 'fantastical experiment' of cloning higher organisms.

1944
Oswald Avery discovers genetic information is carried by the nucleic acids of cells.

1952
Briggs and King clone tadpoles.

1953
Watson and Crick find the structure of DNA.

1958
F.C. Steward grows whole carrot plants from carrot root cells.

Modern years

1962
John Gurdon claims to have cloned frogs from adult cells.

1963
J.B.S. Haldane coins the term 'clone'.

1964
Establishment of the complete genetic code.

1967
Enzyme DNA ligase isolated.

1969
Shapiero and Beckwith isolate the first gene.

1970
First restriction enzyme isolated.

1972
Paul Berg creates the first recombinant DNA molecules.

1973
Cohen and Boyer create first recombinant DNA organisms.

1977
Karl Illmensee claims to have created mice with only one parent.

1978
The release of David Rorvik's book, *In His Image: The Cloning of a Man*, sparks a worldwide debate on cloning ethics.

1979
Karl Illmensee makes claim to have cloned three mice.

1980
US Supreme Court rules live, human made organisms are patentable material.

1983
Kary B. Mullis develops the polymerase chain reaction technique for rapid DNA synthesis.
Solter and McGrath fuse a mouse embryo cell with an egg without a nucleus, but fail to clone using their technique.

1984
Steen Willadsen clones sheep from embryo cells.

1985
Steen Willadsen joins Grenada Genetics to commercially clone cattle.

1986
Steen Willadsen clones cattle from differentiated cells.
First, Prather, and Eyestone clone a cow from embryo cells.

1990
Human Genome Project begins.

1995
Wilmut and Campbell clone sheep from differentiated cells.

1996
Dolly, the first animal cloned from adult cells, is born.

1997
President Bill Clinton proposes a five-year moratorium on cloning.
Richard Seed announces his plans to clone a human.
Wilmut and Campbell create Polly, a cloned sheep with an inserted human gene.

1998
Teruhiko Wakayama creates three generations of genetically identical cloned mice.

- The above information is an extract from the *Conceiving a clone* web site: library. advanced.org/24355/ data/ All research for and development on the web site was done by Kayvon Fata-halian, Bennet Schneider and Brandon Reavis of Kayotic Development

© Kayotic Development
October, 1999

How far should clone research go?

With the cloning of human embryos within reach, Roger Highfield asks who is marking out the ethical boundaries

Advances in cloning are now coming so rapidly that a fresh debate is required to prevent abuses of the science, the chairman of the World Medical Association said yesterday.

The science has made huge strides since the first furore over human cloning in 1993, when researchers at the George Washington University Medical Center reported that they had reproduced in the laboratory the process that leads to identical twins.

Now several teams, including those in Britain, hope to clone human embryos using two well-established methods, developed by the Roslin Institute, Edinburgh, and the 'Honolulu technique' announced last year by the University of Hawaii. 'Our position is that cloning of human beings should not take place,' said Dr Anders Milton, chairman of the WMA. 'But we have to study this further to see where we should draw the line.'

The prospect that human embryos – not people – will be routinely cloned has drawn closer because of efforts to develop commercial treatments for a range of ailments, from diabetes to heart disease, by using cloning to generate a patient's own cells and tissue. Current research plans to dismantle early cloned human embryos to grow tissues for transplant and organ repair, rather than implant the cloned embryos into a surrogate mother to produce a cloned human. Critics argue that such work marks an important step towards the first cloned baby.

Last year, the WMA's meeting reaffirmed its 1997 resolution calling

on doctors engaged in research to abstain from participating in the cloning of human beings. But a new debate is now necessary. The issue is 'of such critical potential consequences affecting the future of medicine and society that we need to make absolutely sure we have considered all the major ethical issues'.

The WMA, which represents 70 national medical associations, including the BMA, will hold a

> *The issue is 'of such critical potential consequences affecting the future of medicine and society that we need to make absolutely sure we have considered all the major ethical issues'*

scientific session to discuss cloning at its annual general assembly meeting in Tel Aviv, Israel, in October.

Among the companies racing to use cloning to make tissue for transplant is the Californian company Geron, which recently moved to pool its technology with that of the Roslin Institute, where Dolly the sheep was cloned three years ago.

Both Roslin and Geron have stressed that they have no intention of cloning human beings. But they do hope to produce human embryos as a source of cells and transplant tissue.

First, the scientists would transfer the nucleus of a cell – taken from a patient that needs tissue – into an unfertilised egg that has had its own nucleus removed. The resulting 'reprogrammed' egg is given a shock of electricity to persuade it to develop into an embryo.

If the Government follows the advice of the Human Fertilisation and Embryology Authority, scientists could use the cloned embryo, up to the 14-day limit allowed for research, as a source of 'stem cells', which can develop into any kind of tissue, whether nerve cells or muscle tissue to repair a damaged heart, and which will not be rejected by the patient.

One team that plans to conduct this work is run by Dr Austin Smith, of Edinburgh University's Centre for Genome Research. He already has an HFEA licence to use spare IVF embryos to study and grow stem cells, which develop in a six-day-old embryo consisting of around 100 cells.

The work is allowed under present legislation because it will help to improve IVF methods, one of five specified areas where embryo research is permitted. But, if the Government approves, Dr Smith said

he would like to use cloning to generate stem cells. And he added that several other British groups were interested 'though they don't want to put their heads above the parapet'.

A survey conducted last year for the Wellcome Trust found that people had reservations about cloning for medical treatments because it involved the destruction of human embryos.

One attempt to sidestep the creation of human embryos was reported last November by the company Advanced Cell Technology, a biotechnology company based in Worcester, Massachusetts, where scientists claim to have created a variation on a human clone by fusing the nucleus of a human cell with a cow egg.

While some will find this cow-human fusion offensive, and critics raise technical issues regarding whether such a hybrid could produce useful stem cells, the technique is regarded by others as an ethical escape route from more offensive ways of producing stem cells, such as cloning a human embryo. And cow eggs are much more plentiful than human eggs.

Many scientists argue that, when the benefits of using early embryos to generate a supply of cells and tissue are clear, the 'yuk factor' will disappear, just as it did after the first heart transplant or first test-tube baby.

Breeding supergods

Professor Lee Silver, one of the world's foremost authorities on genetics and cloning, here paints a chilling picture of the brave new world we are building

Earlier this month, it was reported that the Chicago physicist Richard Seed is poised to begin experiments on cloning human beings, much in the way that Dolly the sheep was cloned from another adult animal, without any sexual involvement.

And now we learn that British Government advisers have recommended paving the way for human cloning by allowing cloned embryos to grow for two weeks in experiments.

Inevitably, there will be expressions of distaste and protest; and of course such experiments involve the deepest moral and ethical questions. But, sadly, I believe objectors are wasting their breath. Cloning, genetic engineering and all the rest of it are now an inevitable part of human progress.

For now, the emphasis may be on helping infertile couples to have children of their own and on stamping out cruel diseases like cystic fibrosis and sickle cell anaemia. But there are far more dazzling prospects ahead.

We humans have, for the first time, the ability to evolve ourselves. No longer must we rely on the slow

processes of generation over millions of years. We can now choose what genes we want to put into our children; and we can do it right now.

What it means is that in the next 500 years – and that's a conservative estimate – we can, and almost certainly will, adapt ourselves far more dramatically than anything evolution has accomplished in the past five million.

Even now in the United States we have embryo screening, through which parents who can afford to pay ensure that their unborn children are free of disease. Embryos which show signs of damage are routinely terminated.

So what will happen when the gene which predisposes a human being to alcoholism is found? Or the gene which influences obesity? Or aggressiveness? It would be only the smallest of steps to eliminate such traits too.

The time will come – sooner than most members of the public think – when we shall not only be able to identify damaged or disease-prone genes, but the elements in the human make-up which determine talent, athletic ability, good looks and artistic skill.

The consequence is inevitable. Parents will want to give their children one or more of those special qualities. Of course it won't necessarily mean those children will turn out to be Olympic athletes or concert pianists – that takes study, determination and sheer persistence, too – but it will give them a flying start in life.

And who can stop it happening? After all, what politician would stop a mother using the methods of genetic engineering to ensure her unborn child is free of heart disease?

And once that principle is accepted, where can the line be drawn? Who is qualified to decide that parents must not determine the sex, colouring and physical characteristics of their own children?

And here's where I see a nightmare developing.

One day, probably within the next 50 years, we will be able to isolate the gene which determines intelligence. Once again, market forces and the natural desire of couples to secure the best for their children will take over. Those who can afford the price will certainly pay if it means their offspring will turn out brighter than the rest.

Once the technique is established, prospective parents – invariably the better-off – will seek the specialist agencies who can guarantee genes that will provide intellectual brilliance.

Intelligence for the fortunate minority will be added on, generation after generation. The present class system is fairly flexible because it is based on economic circumstances which can change. But in future I foresee a genetic class system, with the gulf between the genetically enhanced 'superbeings' and the rest becoming unbridgeably wide.

I'll go further. In time – probably a matter of a few generations – those two classes of human being will become virtually separate species. They won't intermingle except possibly in the relationship of master to servant. They won't intermarry. They certainly won't breed across the intellectual divide.

Indeed I foresee a whole new range of social and ethical problems in the new dispensation. For example: how will the superbeings treat their ordinary cousins? As slaves? Or as enemies?

Some may see all this as mere fantasy. But I truly don't think it is. Five million years ago, humans and chimpanzees were indistinguishable. Even today, we share 99 per cent of the same genetic make-up. The difference between our two species lies in that crucial 1 per cent.

When we've identified and understood just how that 1 per cent difference works – and when we've applied what we have learned to human intelligence – the potential is mind-boggling.

In time, for example, we shall probably be able to defeat death. We are all programmed to die; but when we've identified the precise mechanisms which cause ageing and death we will be able to alter them. There is no reason why human beings shouldn't live to be 200 or 300.

In the future I foresee a genetic class system, with the gulf between the genetically enhanced 'superbeings' and the rest becoming unbridgeably wide

---

The future holds other wonders. With every generation, more genetic enhancements can be added to the human frame. A human species will develop which is not only brighter, healthier and longer-lived, but which can incorporate any of the special talents of the animal kingdom.

There is no scientific reason why humans shouldn't incorporate the acute sense of smell which dogs enjoy; or why they shouldn't acquire the sonar ability of bats, or the sensitivity to magnetic forces which birds use to navigate across continents.

In fact, there is no characteristic of any animal which might not one day be acquired.

It could be argued then that men – the wealthiest of them at any rate – will have turned themselves into gods. Some, no doubt, will react with revulsion at the prospect of Man tampering so fundamentally with nature itself.

Yes, I know there are sceptics who say a thing can never happen. There always are.

In 1935 they told us that understanding the true nature of the gene was beyond the power of mortal man. In 1974 they said it was impossible to determine their sequence. In 1984 they said it was impossible to alter specific genes within the embryo. In 1996 they insisted it was impossible to clone people from adult cells.

The lesson we should have learned by now is that the pace of scientific progress is so rapid that today's miracle is tomorrow's routine. In embryology, we are well embarked on a new age of miracles.

• The above information is an excerpt from *Remaking Eden: Cloning And Beyond In A Brave New World* by Lee Silver (Weidenfeld & Nicolson, £20).

© Lee Silver, 1998

Human cloning

Your questions answered

Q. What is cloning?
A. Human cloning is the creation of a human being, or a number of human beings, who is/are genetically identical to another.

In the current debate, you are likely to hear cloning referred to in one of two ways:
• *Reproductive cloning* refers to the creation of a new person with the same genetic make-up as someone who is alive or has lived.

• *Therapeutic cloning* refers to using cloning techniques to initiate the growth of embryos in order to create new organs or cells for medical and research purposes.

Q. How does cloning occur?
A. Artificial cloning can be carried out by one of two techniques:
• *Embryo splitting* is similar to the natural process which creates identical twins – the embryo's cells are separated at a very early stage of development to create one or more clones.
• *Nuclear replacement* is the process that was used to create 'Dolly' the sheep.[1] It works by taking a cell nucleus from one person (e.g. from a skin cell) and putting it into the egg of another, whose nucleus has already been removed. The egg is then stimulated to divide, for example, by treatment with

bursts of electric current, thus starting the growth of an embryo. The egg's nucleus contains all the chromosomes (genetic information) for the cell to start dividing and become a fully developed child (or lamb!) in due course.

Nuclear replacement technology can be used to create a clone from a person of any age (embryo, foetus or adult) and it can potentially create many more clones than embryo splitting.

Q. Does UK law currently allow human cloning?

A. Yes and No. The law currently does not allow any form of reproductive cloning to be licensed, but there is no explicit ban on cloning. The Government have made it clear that they believe 'human reproductive cloning is ethically unacceptable and cannot take place in this country'.[2] This is a very welcome statement.

Nuclear replacement of an embryo, or any cell that is part of any embryo, is also prohibited.

Nuclear replacement of an egg (like the process that led to Dolly the sheep) and embryo splitting are currently allowed in human embryos under the jurisdic-tion of the Human Fertilisation and Embryology Authority (HFEA) only for short-term experimental use. However, the HFEA 'has made it crystal clear that it will not license cloning by embryo splitting for treatment purposes'.[3] The HFEA also regulates the legislation which allows embryos up to 14 days old to be destroyed, frozen or researched on for specific purposes (laid out in the Human Fertilisation and Embryology Act 1990).

There would need to be an amendment to the Act to allow the HFEA to license embryo research into 'therapeutic approaches to disease or tissue damage',[4] i.e. extending the Act for a new purpose. However, prior to the HFEA Act being in place, the Warnock Report of 1984 made it clear that human cloning should not be permitted, so any change in this policy would be contrary to the spirit of the law.

Discussion about the law on cloning has not been restricted to the UK.

- In March 1997, the European Parliament voted in a resolution to ban cloning of human beings. However, the resolution carries no legal weight. This policy area remains more obviously a matter for individual member states, with national politicians bringing in legislation on human cloning.
- In December 1997, UNESCO published the Universal Declaration on the Human Genome and Human Rights. Article 11 bans 'practices which are contrary to human dignity, such as reproductive cloning of human beings . . .'.
- In Strasbourg, the Council of Europe, with 40 member states, produced a protocol banning human cloning which has been added to their Convention on Biomedicine and Human Rights, and was signed on 12 January 1998 by 19 member states. However the UK has not yet ratified this Convention so cannot sign the Protocol.

Q. Why might people want to use cloning?

A. Cloning seems to offer new possibilities of medical advances. Some people might want to use cloning to replace a dead baby or child; or indeed have a child if they are infertile. Others may want to replicate themselves for future generations, while yet others may want to create a clone to produce an organ for transplantation without the complications of rejecting the organ, i.e. for therapeutic purposes.

References

1 The description 'somatic-cell nuclear transfer' is the term used in the USA.
2 Reply to House of Lords Written Parliamentary Question, 24 June 1999, col 106.
3 Baroness Hayman, House of Lords debate on Human Embryos and Cloning, 28 April 1999, col 350.
4 *Cloning Issues in Reproduction, Science and Medicine*, December 1998, para 5.9

• The above information is an extract from *Human Cloning: Your Questions Answered*, a paper published by the Public Policy Department of the Christian Action Research Council.

© Christian Action Research and Education (CARE)

Cloning misconceptions

Information from *Conceiving a Clone* – Kayotic Development

Most cloning misconceptions arise from a lack of knowledge. Most people do not understand the basic principles of cloning, and are likely to make rash generalisations about whether cloning is natural or not. Other misconceptions focus on the societal problems resulting from cloning. Many of these misconceptions are only valid in a society without regulations or laws of any kind. People forget that along with new technological developments come rules and guidelines to prevent the kind of scenarios here. Each misconception results from a distortion of the truth, which is presented here with each incorrect belief.

A clone would not be a normal human

Whatever the methods of production are, a clone would be as 'human' as an identical twin. Both are derived from a single fertilised egg.

Cloning is 'playing God'

Cloning does not create life, as this stigma implies. Cloning merely produces life from existing life. Cloning can be thought of as an extension of procedures like in-vitro fertilisation.

Cloning is not a natural process

Cloning utilises elements that already exist in the natural reproduction process. Embryo cloning pulls apart a zygote at the two-cell stage and creates two one-celled organisms. Although some might say that cloning is not an intended form of reproduction, the same might be said of in-vitro fertilisation, and the use of fertility drugs.

A clone will not have a soul

This implies that the soul is a quantifiable physical element of someone's genetic make-up that can be altered or taken away. In this case, cloning does not present more of a religious problem than identical twins. Despite them being identical, it is agreed that both twins have souls.

A clone will have the same feelings and emotions as its genetic parent

An overused example of this idea is a Hitler clone starting a new Holocaust. While genes and genetic structure can give certain characteristics and possibly basic emotional tendencies, environment and upbringing play a much larger role in shaping someone's emotions and outlook. A Hitler clone that had been raised in the United States and had lived in a period of stability and prosperity would not act the same way as a Hitler raised in Germany living amongst post-war devastation and hatred.

An unconscious clone could be produced to supply organs

Despite being morally questionable, producing a clone with no self-awareness requires a deeper knowledge of where the consciousness resides. Consciousness is not a certain trait that can be erased through genetics, and there is no isolated DNA that determines its existence. Furthermore, just proving that a clone is not self-aware would be difficult. People with debilitating neurological disorders may appear mentally incapacitated but retain full consciousness. However, researchers have theorised the possibility of cloning only certain organs to use as replacements for an individual in dire need of a transplant. Scientists believe that if the cells of an organ have the same genetic make-up as those of the host organism, the organ would be much less likely to be rejected after a transplant.

Someone could own a clone

Although cloning is being considered as a future infertility remedy, and essentially, a clone would be 'made' for the parents, no one could own a clone. Ownership of a clone would be no different than slavery. People that predict a massive working underclass produced from cloning forget that despite the methods of their birth, clones would carry the same rights as a person produced through normal reproduction.

Great individuals of the past could be reborn

All current techniques to clone an adult cell use the method of nuclear transfer, which requires the donor cell to be alive. In this process a LIVE adult cell is fused with an egg cell or its nucleus is extracted and inserted into the egg. At this time, and most likely far into the future, clones of dead organisms cannot be created. Also, even if such an individual is cloned, the development of the person is largely dependent upon its upbringing and childhood surroundings. Just as a theoretical Hitler clone would most likely not grow up to start a new Holocaust, an Einstein clone would probably not become a world-renowned physicist.

• The above information is an extract from the *Conceiving a clone* web site: library.advanced.org/24355/data/ All research for and development on the web site was done by Kayvon Fatahalian, Bennet Schneider and Brandon Reavis of Kayotic Development

© Kayotic Development

> *Most people do not understand the basic principles of cloning, and are likely to make rash generalisations about whether cloning is natural or not*

Pros and cons of cloning

Editorial comment

Welcome to Brave New World. Consider this. It is now possible for scientists to remove the nucleus (containing life's genetic formulae) from one of your body cells; insert it into an unfertilised human egg that has had its own nucleus removed; jerk the 'reprogrammed' egg into life with an electric shock; and, voila, you have been cloned . . .

Once created, your embryo clone could grow into a human being. He/she (it?) would be the closest relative you have known. If a he, would he be your son or brother?

This is frightening stuff, never mind the insouciance of researchers who are anxious to get their tweezers into mass-produced embryos. The first known human embryo cloning – by South Korean scientists last December – marked a quantum leap from man as created by God to God-like man. We entered uncharted ethical territory.

The prospect excites many while filling others with foreboding. Professor Vivian Nathanson, head of policy research at the British Medical Association, envisages the cloning of an embryo programmed genetically, once it grows, to produce bone marrow to be donated to a child already ill from leukaemia. The 'marrow-healthy' embryo would be implanted into the womb of an infertile mother distressed that she has been unable to bear a child.

Happiness and well-being all round. Indeed, Scotland's Roslin Institute, where Dolly the sheep was cloned three years ago, has entered a deal with California's Geron Corporation, a world leader in genetic research, to clone human embryos which would be broken up and 'bio-pharmed' to produce human skin, cartilage, hearts, spleens, livers and neurones for transplant into diseased humans.

To eugenicists who were so fashionable before the Second World War – such as Sweden's chillingly titled National Institute for Racial Hygiene (NIRH), wound up after sterilising 60,000 misfits who did not meet the ideal of a blond, blue-eyed, intelligent Scandinavian – it would be bliss to be alive today. Human embryo cloning research is permitted in the private sector in the US and other countries. A human embryo was cloned this month by the Advanced Cell Technology Corporation in Massachusetts and has since been destroyed.

> *The battle lines are drawn, but without clarity. As so often with science we seek the fine balance between greater good and lesser evil*

Apparently rational, but ultimately immoral, scientists of the kind who worked for 40 years in the NIRH lie at the root of objections to human embryo cloning. Objectors say it is wishful thinking to suggest that some for-profit clinic in the mid-West will not pop up with the first cloned human being in breach of laws that demand that cloned embryos be destroyed: profound moral and ethical questions are raised when men tamper with the very stuff of life. Of scientists, objectors mimic the barbed aphorism Churchill applied to Stafford Cripps: 'There, but for the grace of God, goes God.' They argue that the ultimate horror has arrived when man usurps the role of both God and nature and becomes the arbiter of his own creation, disrespecting both the dignity and mystery of human life. Tough scientists dismiss this religious position on the sanctity of human life as an amalgam of credulousness, bigotry and false deductions based on myth.

The battle lines are drawn, but without clarity. As so often with science we seek the fine balance between greater good and lesser evil. We must aim to restrain the latent arrogance of science without denying hope to those for whom genetic science is, in the best sense, a real and actual miracle. The Government is right to delay for another six months a decision on whether to permit in Britain the creation of cloned human embryos for experimental purposes, but a very great deal more public debate is necessary.

© *The Scotsman*
June, 1999

The benefits of human cloning

Information from the Human Cloning Foundation

There are many ways in which human cloning is expected to benefit mankind. Below is a list that is far from complete.

Rejuvenation

Dr Richard Seed, one of the leading proponents of human cloning technology, suggests that it may someday be possible to reverse the ageing process because of what we learn from cloning

Human cloning technology could be used to reverse heart attacks

Scientists believe that they may be able to treat heart-attack victims by cloning their healthy heart cells and injecting them into the areas of the heart that have been damaged. Heart disease is the number one killer in the United States and several other industrialised countries.

There has been a breakthrough with human stem cells

Embryonic stem cells can be grown to produce organs or tissues to repair or replace damaged ones. Skin for burn victims, brain cells for the brain damaged, spinal cord cells for quadriplegics and paraplegics, hearts, lungs, livers, and kidneys could be produced. By combining this technology with human cloning technology it may be possible to produce needed tissue for suffering people that will be free of rejection by their immune systems. Conditions such as Alzheimer's disease, Parkinson's disease, diabetes, heart failure, degenerative joint disease, and other problems may be made curable if human cloning and its technology are not banned.

Infertility

With cloning, infertile couples could have children. Despite getting a fair amount of publicity in the news current treatments for infertility, in terms of percentages, are not very successful. One estimate is that current infertility treatments are less than 10 per cent successful. Couples go through physically and emotionally painful procedures for a small chance of having children. Many couples run out of time and money without successfully having children. Human cloning could make it possible for many more infertile couples to have children than ever before.

Plastic, reconstructive, and cosmetic surgery

Because of human cloning and its technology the days of silicone breast implants and other cosmetic procedures that may cause immune disease should soon be over. With the new technology, instead of using materials foreign to the body for such procedures, doctors will be able to manufacture bone, fat, connective tissue, or cartilage that matches the patient's tissues exactly. Anyone will be able to have their appearance altered to their satisfaction without the leaking of silicone gel into their bodies or the other problems that occur with present-day plastic surgery. Victims of terrible accidents that deform the face should now be able to have their features repaired with new, safer, technology. Limbs for amputees may be able to be re-generated.

Breast implants

Most people are aware of the breast implant fiasco in which hundreds of thousands of women received silicone breast implants for cosmetic reasons. Many came to believe that the implants were making them ill with diseases of their immune systems. With human cloning and its technology breast augmentation and other forms of cosmetic surgery could be done with implants that would not be any different from the person's normal tissues.

Defective genes

The average person carries eight defective genes inside them. These defective genes allow people to become sick when they would otherwise remain healthy. With human cloning and its technology it may be possible to ensure that we no longer suffer because of our defective genes.

Down's syndrome

Those women at high risk for Down's syndrome can avoid that risk by cloning.

Tay-Sachs disease

This is an autosomal recessive genetic disorder that could be prevented by using cloning to ensure that a child does not express the gene for the disorder

Liver failure

We may be able to clone livers for liver transplants

Kidney failure

We may be able to clone kidneys for kidney transplants

Leukaemia

We should be able to clone the bone marrow for children and adults suffering from leukaemia. This is expected to be one of the first benefits to come from cloning technology.

Cancer

We may learn how to switch cells on and off through

cloning and thus be able to cure cancer. Scientists still do not know exactly how cells differentiate into specific kinds of tissue, nor do they understand why cancerous cells lose their differentiation. Cloning, at long last, may be the key to understanding differentiation and cancer.

Cystic fibrosis
We may be able to produce effective genetic therapy against cystic fibrosis. Ian Wilmut and colleagues are already working on this problem.

Spinal cord injury
We may learn to grow nerves or the spinal cord back again when they are injured. Quadriplegics might be able to get out of their wheelchairs and walk again. Christopher Reeves, the man who played Superman, might be able to walk again.

Testing for genetic disease
Cloning technology can be used to test for and perhaps cure genetic diseases.

The above list only scratches the surface of what human cloning technology can do for mankind. The suffering that can be relieved is staggering. This new technology heralds a new era of unparalleled advancement in medicine if people will release their fears and let the benefits begin. Why should another child die from leukaemia when if the technology is allowed we should be able to cure it in a few years' time?

From e-mail to the Human Cloning Foundation it is clear that many people would support human cloning in the following situations:

1) A couple has one child then they become infertile and cannot have more children. Cloning would enable such a couple to have a second child, perhaps a younger twin of the child they already have.

2) A child is lost soon after birth to a tragic accident. Many parents have written to the HCF after losing a baby in a fire, car accident, or other unavoidable disaster. These grief-stricken parents often say that they would like to have their perfect baby back. Human cloning would allow such parents to have a twin of their lost baby, but it would be like other twins, a unique individual and not a carbon copy of the child that was lost under heartbreaking circumstances.

3) A woman who through some medical emergency ended up having a hysterectomy before being married or having children. Such women have been stripped of their ability to have children. These women need a surrogate mother to have a child of their own DNA, which can be done either by human cloning or by in-vitro fertilisation.

4) A boy graduates from high school at age 18. He goes to a pool party to celebrate. He confuses the deep end and shallow end and dives head first into the pool, breaking his neck and becoming a quadriplegic. At age 19 he has his first urinary tract infection because of an indwelling urinary catheter and continues to suffer from them for the rest of his life. At age 20 he comes down with herpes zoster of the trigeminal nerve. He suffers chronic unbearable pain. At age 21 he inherits a 10-million-dollar trust fund. He never marries or has children. At age 40 after hearing about Dolly being a clone, he changes his will and has his DNA stored for future human cloning. His future mother will be awarded one million dollars to have him and raise him. His DNA clone will inherit a trust fund. He leaves five million to spinal cord research. He dies feeling that although he was robbed of normal life, his twin/clone will lead a better life.

5) Two parents have a baby boy. Unfortunately the baby has muscular dystrophy. They have another child and it's another boy with muscular dystrophy. They decide not to have any more children. Each boy has over 20 operations as doctors attempt to keep them healthy and mobile. Both boys die as teenagers. The childless parents donate their estate to curing muscular dystrophy and to having their boys cloned when medical science advances enough so that their DNA can live again, but free of muscular dystrophy.

• The above information is from the Human Cloning Foundation web site which can be found at www.humancloning.org/

Government rejects advice and blocks human cloning

The Government yesterday rejected expert advice and blocked any move to clone human embryos for the production of tissue for transplants and a range of other medical treatments.

Scientists, who had been expecting therapeutic cloning to be approved, said the decision would mean Britain lagging behind in an area of research in which it excelled and accused the Government of giving in to media scaremongering.

The Government also reaffirmed in the House of Commons yesterday that reproductive cloning – the cloning of people – was 'ethically unacceptable' and could not be carried out in this country. It was responding to a report last December on cloning by the Human Fertilisation and Embryology Authority and the Human Genetics Advisory Commission that recommended cloning embryos for the supply of cells and tissue for medicine. Unexpectedly, the Government rejected this advice and said that more evidence was needed – from animal research, for instance – of the potential benefits to human health.

Therapeutic cloning would harness the techniques used to clone Dolly the sheep to turn a cell from a patient into compatible tissue and cells for transplant. Controversially, this would require the destruction of an early cloned embryo.

Tessa Jowell, minister for public health, and Lord Sainsbury, the science minister, announced that an independent, expert advisory group was to be set up to assess therapeutic cloning. The group will be chaired by Prof Liam Donaldson, Chief Medical Officer. He said yesterday that the new group – consisting of 'the best brains in this country and overseas' – would report its findings early next year. Until then there can be

By Roger Highfield, Science Editor

no chance of a change to the 1990 Human Fertilisation and Embryology Act to allow human cloning for therapeutic purposes.

'There is a lot of public concern about this field and it is important we proceed very carefully,' he said. 'We are, for the moment, not allowing such research to be carried out. It's up to you if you want to call that a moratorium.'

Prof Donaldson dismissed the suggestion of top scientists leaving Britain to do research abroad, as the advisory group would report in a relatively short time.

> **The Government rejected advice and said that more evidence was needed – from animal research, for instance – of the potential benefits to human health**

The head of the spin-off company from the Roslin Institute, where Dolly was cloned, welcomed the ban on reproductive cloning but said yesterday that Britain could now be left behind in the race to conduct therapeutic cloning. 'We are disappointed that the Government is not willing to endorse the recommendation to allow cloning for stem cell research,' said Simon Best, managing director of Geron Bio-Med. The company was formed from the recent merger of Roslin Bio-Med with the Californian biotech company Geron, which is already conducting human experiments in America.

'Research groups in this country will be stuck and there is a real risk that Britain is going to be left behind,' said Mr Best yesterday. 'This is not just about replacing existing treatments. It's a whole new market which could benefit the economy and an area in which we are very smart at the moment.' He added that the new advisory body would simply go over the same territory as the December report.

Prof David Latchman, vice-chairman of the Parkinson's Disease Society's medical advisory panel, accused the Government of being over-cautious and driven by newspaper headlines. 'I think this is a fall-out from GM food, which in itself was a fall-out from BSE.'

He said evidence of the need for such new treatments was only too visible in hospital wards around the country. 'There are huge numbers of very desperate people out there.'

The shadow health secretary, Dr Liam Fox, said the Government had shown itself incapable of making a decision over the issue of human embryo cloning, preferring once again to hand the responsibility over to scientific advisers. He added: 'We elect governments to govern us, not scientists, and it is therefore the job of government to have the courage to take difficult decisions, but which this Labour Government seems incapable of grasping.'

The Christian think-tank Care – Christian Action Research & Education – said it was delighted that the Government had rejected the advisory bodies' recommendation. 'We urge them to ensure that the new review body considers not just the science but the gaping moral black hole of such scientific research,' said a spokesman. 'It is also essential that the new body is purged of any corporate or commercial interest by the big pharmaceutical companies.'

Forces that put cloning under a shadow

Researchers confident that despite a hiatus, human tissue will be grown within 10 years

By Tim Radford, Science Editor

Most scientists in human genetic research in Britain took the philosophical view last night. A government frightened by public resentment of genetically modified foods was hardly likely to push its luck with something that would be presented as approval for human cloning.

So, despite a year of contemplation, some very searching questions and approval from two government-appointed committees, the research goes on hold. There will be no human clones. And there will be no use of embryo stem cells to clone human tissue, at least for a while.

But researchers yesterday were privately confident that, within 10 years, medical laboratories would be 'growing' human tissue to help combat heart disease and neuro-degenerative diseases such as Huntington's chorea and Parkinson's. The argument is that the technique is out there, already being examined in commercial laboratories in the US and elsewhere – and too powerful a treatment to ignore for long.

Treatment in 10 years

'My parents are in their early 70s,' said Simon Best of Roslin BioMed, one of the new biotechnology companies at the forefront. 'By the time they are in their early 80s it might help them. Right now we know from animals that we can develop heart muscle cells and neurons for Parkinson's and perhaps some of the blood precursors relatively quickly. Those technically are likely to be the easiest and might get into the clinic in five to seven years and might be available to patients within a 10-year time frame.'

Humans clone bits of themselves all the time: it is called growth and renewal, or healing. Sometimes the process comes to a halt in one place or another, and unless surgeons resort to a heart or liver or bone marrow transplant, death follows. But transplants depend on somebody else dying while in possession of a donor card – and there are too few donors.

Transplants have been tried from animals – but the patient then has to live with a chemically suppressed immune system. The ideal would be to grow 'personalised' new human organs which could be used to replace failing tissue: to take samples from the patients and cultivate new flesh in a laboratory dish. Until 1997, nobody knew how to grow new organs without growing a new human.

Then the Roslin Institute in Scotland cloned Dolly the sheep from an adult cell. And in 1998, the Geron corporation in the US – one of the new biotechnology companies which have made the human body their business – discovered how to preserve and cultivate human embryo stem cells. These are the master cells, the little fountains of life: they are the cells that 'make' blood and bone and skin and nerve cells.

Suddenly the research had somewhere to go. In Britain, there were people who knew how to transfer the DNA from one animal into the cells of another. In America there were people who knew how to manipulate the way cells developed. It would be possible to pool the technologies and tackle diseases for which there is still no treatment, and only the dimmest hopes of one.

But 'therapeutic' cloning is not like human cloning. Both techniques start the same way: with DNA being added to an egg, and with the egg being triggered into dividing to

embark on the road towards the status of an embryo. The therapeutic cloners would start using cells at a very early stage – long before it would begin to develop into something recognised as a foetus.

But once again, this knowledge raised huge questions about the nature of life, and the responsibilities of those who manipulate it. They have already been answered in part. Infertility treatments have involved embryo research for more than two decades – but that was the use of embryos to create a human life.

Therapeutic cloning seems to involve the handling of human embryos to prevent death, a different matter.

The scientists at the leading edge knew that the decision would be for society as a whole.

But it may not work out so neatly. In the US, there is no federal funding for human embryo research – but there are no restrictions on what private biotechnology com-panies may do. In Britain, both private and government researchers have to apply to government agencies for licences to do carefully specified research on embryos.

Austin Smith, of the Centre for Genome Research in Edinburgh, yesterday welcomed the delay, but hoped it would not be permanent. He wants to see such research conducted in public, and scrutinised by the public.

But commercial enterprises are poised, too. Earlier this year, Roslin BioMed – a business built on the technology that led to Dolly the sheep – joined forces with the Geron corporation in the US. Simon Best, of Roslin BioMed, was disappointed, because he had hoped, after a year of contemplation, for a government go-ahead. 'Britain is at the forefront of a lot of bits of this technology.

'But there are a lot of competi-tors in the US. The delay could have some negative consequences,' he said.

Technique's big potential

- A clone results from the successful transfer of adult DNA into a new egg. Sheep, cows, and mice have been copied, but success is still relatively rare.
- Cloning could be used to 'recreate' prize-winning breeding cattle, or to 'copy' genetically engineered animals which produce valuable human proteins such as human albumen or human clotting agents.
- Since a clone's DNA would be as 'old' as that of its parent, it might be useful for studying the mutations that lead to increased risk of cancer with age.
- Cloning might be the only way of saving species balanced on the edge of oblivion. The Chinese have suggested cloning the giant panda.
- The most immediate hope is to use the technique to treat diseases that have no foreseeable cure.

© The Guardian
June, 1999

The clone guinea pigs?

Infertile couple volunteer to have the first carbon-copy baby

A British couple have volun-teered to parent the world's first cloned baby.

Peter and Ildiko Blackburn, described as wealthy and intelligent, say they have thoroughly researched the issues and are ready to be the first guinea pigs if the process becomes legal.

In the wake of Dolly the sheep, the science of cloning has been progressing at a breathtaking pace. South Korean researchers already claim to have created and destroyed the first cloned human embryo.

Cambridge graduate Mr Blackburn, 31, and his 29-year-old wife have been trying for a child without success for more than half of their eight-year marriage, undergoing extensive fertility treatment.

The couple, from Huntingdon, Cambridgeshire, say they are willing to take their place in history and have his skin or blood DNA fused with one of his wife's eggs and implanted back into her womb. If the baby survived it would be an exact replica of its father.

Speaking on BBC1's *Panorama* programme, Mr Blackburn argues that a cloned child would be no different to one naturally conceived and would still be an individual, shaped by the social and economic circumstances into which it was born.

'I'm not comfortable with the idea of there being 10,000 little mes, he says. 'But a child is a child. It would not be treated any differently and it would be loved.'

Cloning babies is banned in Britain under the 1990 Human Fertilisation and Embryology Act but the Government is expected to re-examine its position in the light of recent developments.

In December, a team of South Korean researchers claimed to have created the first cloned human embryo but destroyed it within a few days for 'ethical' reasons.

Dr Lee Bo Yon, the scientist responsible, admits that the egg was taken from a woman who had no knowledge of what it would be used for. 'We were not in the advanced cloning stage and were afraid in case the patients asked for the egg to be put in the womb if we mentioned the word cloning,' he said. But he admits his work is continuing.

The developments have deeply concerned opponents of cloning. Campaigner Lord Alton said last night: 'What is it in us, what enormous sort of conceit, that wants to create more of ourselves in a sort of conveyor belt? Diversity is important and we interfere with this process at our peril.'

Dr Ian Wilmut, who headed the team which created Dolly the sheep, believes human cloning is unethical. 'I have not heard a reason for human cloning with which I am personally comfortable,' he said.

© The Daily Mail
February, 1999

Human cloning

Should it be done? What would it mean? Frequently asked questions

Until the birth of Dolly, it was believed that the ability to clone an adult human was either impossible or possible only in the distant future.

However, human embryos have been 'twinned' in the past! These were not true clones (although they have been called clones) and they were not viable, however, for they did not survive.

What does human cloning 'mean'?

The emergence of new technologies creates a new set of cultural events and their consequences with which human cultures must come to terms. Humans must define a status and role for any new technology. This is a process of adaptation and acculturation. In a sense, this is a process whereby humans try to decide what a new technology 'means' to them. Questions frequently asked during this process of acculturation and adaptation include:

- What are the needs and goals this technology might serve? Is this technology the only means for addressing these goals and needs?
- What members of which communities have these goals and needs?
- Of these groups in need, who will have access to the technology and its products; for example, will it be only special individuals or classes of people?
- Who will be benefited, and who will be harmed, indirectly as well as directly, by implementation of this new technology?
- How reliable, how safe, and how well can this new technology be controlled?
- What is the best case, and the worst case, scenario if this technology were encouraged, or if it were impeded, in its development and implementation?
- Is the new technology to be defined as 'necessary' or 'essential', or 'innocuous' or 'superfluous', or 'good' or 'bad'?
- Is the technology, or are its consequences, to be deemed ethical, moral, immoral?
- Should the technology be considered a form of 'property', to be held by individuals or publicly held and administered? And if so, by whom?

What is this new biotechnology?

We are in the process of deciding to define cloning as a kind of science rather than art or religion, specifically a kind of science we call biotechnology . More specifically, we will call this kind of biotechnology 'cloning via nuclear transfer'. Furthermore, cloning can be made distinct from 'twinning'.

Can adult humans be cloned?

Probably yes, and in the very near future. But a great deal more research and development of the nuclear transfer techniques used to clone Dolly is needed. And it must be improved and perfected for use on human embryos. Sheep embryos have some special characteristics that make cloning them much easier than cloning human embryos. Cloning an adult sheep was extremely difficult to do; over 270 attempts were needed before Dolly was born. Many fetal lambs did not survive the early stages of development. Those lambs that were carried to term were born with health problems, including malformed kidneys, and all but Dolly subsequently died. See for example the *Washington Times*, 'Before there was Dolly, there were Disasters' (March 11, 1997).

Should humans be cloned?

A *Time Magazine* poll (March 10, 1997) reported that 74% of those asked believe it is against God's will to clone human beings. President Clinton has banned federal funds from being used for human cloning research, stating that:

'Any discovery that touches upon human creation is not simply a matter of scientific inquiry, it is a matter of morality and spirituality as well ... Each human life is unique, born of a miracle that reaches beyond laboratory science ... '

But others argue in favour of continuing human cloning research, of continuing to clone human embryos and perhaps cloning adult humans in the future. Some arguments in favour of human cloning might include the fact that cloned human embryos would make research into genetics and genetically related diseases, and their treatments or preventions, much easier and cheaper. Cloning embryos could also facilitate the process of in-vitro fertilisation, since the collection and

replacement of ova is often painful and traumatic, and can be unsuccessful.

Embryo cloning is also seen as a potential treatment for infertility when in-vitro fertilisation is not available, such as when parents are infertile, or when one or both parents harbours a genome coding for certain undesirable traits or diseases, or if the parents are homosexual couples. Cloned embryonic tissues might be used for the replacement of lost or diseased tissues.

Adult cloning might appeal to those who desire children/adults who are genetically identical to themselves, or genetically identical to someone whom they love or admire. There may be many other, personal reasons why parents would want their children to be genetically identical to someone who is a non family member. Cloning could provide a genetically identical replacement for a lost loved one.

The belief here is that cloning can be justified as an expression of reproductive freedom of choice, a choice that should not be limited by legislation.

However, it is important to remember that a genetic clone, although sharing an identical genome with their donor, will not be physically and behaviourally identical to their donor! The clone will only be genetically identical to the donor. Their physical and behavioural characteristics will differ in many important and significant ways!

Is human cloning good or bad, moral or immoral, from a religious perspective?

The most commonly cited ethical and moral arguments against human cloning seem to originate from religious perspectives. These religious arguments can even be made by politicians and scientists with religious sympathies. Many religious philosophies teach, for example, that human life is unique and special and should be created, determined and controlled only by their deities. Many religions believe in the existence of, and in the individuality of, a human soul. Many Christians, for example, will be concerned about whether it

will be possible to clone the human soul, along with the human. If it is possible to clone the soul, what will this 'mean'? In contrast, if a person is cloned, but not their soul, what will this 'mean'? Can a clone without a soul be destroyed and not offend moral or religious beliefs? Cloning will be divined by many as humans assuming the powers, the providence, and the jurisdiction of their deities or other spiritual powers of their supernatural universe.

Not all religious leaders feel the same. In contrast to the opinions of their peers, some Jewish and Muslim religious leaders testified before the National Bioethics Advisory Commission that they feel that embryo and cloning research might provide discoveries that would lead to an appropriate way to counter infertility.

Is human cloning legal? Is human cloning scientifically ethical?

Currently, human cloning is illegal in England and Norway, for example, but not illegal in the US. However, in the US, federal, but not private, funds are prohibited from being used to create human embryos (1994) or do research on human embryos if they will be harmed or destroyed (1996-97). In addition, President Clinton has imposed a moratorium on human cloning research (March 4, 1997). Meanwhile, several states in the US have laws restricting embryo research.

As decision makers in the US debate whether or not to support research on human embryos and human cloning many ethical and legal questions arise.

For example, how will the federal ban on human cloning research and the ban on certain types of human embryo research affect other, related fields of research that are deemed important? Human embryo research and embryo cloning can be used to conduct research and development of contraceptives, studies aimed at understanding the causes of human infertility and its solutions, research involving genetic testing, genetic engineering, disease diagnosis, prevention and treatment, and in testing various medicines and medical procedures.

In contrast, if the government

funds this type of research, then it will have some important control over the nature of the research. But what kind of controls might these be? Will the government decide to have an interest in protecting embryos from certain kinds of research? Would unused embryos, left after in-vitro fertilisation procedures, be treated as 'spare' embryos and given a different status for research purposes? Will Democrats vote on this research the same way the Republicans would?

If the federal government decides to continue to not fund human embryo and cloning research, then the government will not have one important avenue for controlling, to some degree, the nature of the research. If the government refuses to support this research, would a funding vacuum be left that market forces will quickly fill?

If the private sector is left to fund research and development, then will this research be driven by entrepreneurial profit motives? What effects will entrepreneurial forces have on the nature of human embryo and cloning research and development?

Is there Constitutional protection for research on human embryos and human cloning? Does the First Amendment guarantee academic freedom, and the right to think, inquire, and do research? When or should the research involving human embryos and cloning be defined as 'academic research and inquiry'? If this research is defined as academic, should the scientific research on human embryos and cloning be protected under the First Amendment, which guarantees the 'freedom of speech'?

There are limits to the kinds of speech that the First Amendment protects. For example, it does not protect speech that is deemed to be obscene, or speech deemed threatening to national security. Some citizens believe that research into and/or trying to clone humans is wrong, while others disagree. When considering whether or not cloning research, or other kinds of academic research, inquiry, and scientific communication are to be protected, and to what degree, by the First Amendment, the government decision makers have to decide how best to balance protecting the freedoms of speech and inquiry of the scientific community – and the benefits their research might produce – against the need to protect other citizens from any dangers this kind of freedom of speech, and its products, might also produce.

In the absence of governmental controls, can/or should the scientific community regulate itself, through peer review, when it comes to human embryo and cloning research? Should society entrust the scientific community to regulate themselves? Would this allow and encourage practices leading to conflicts of interest?

Should some other private organisation, independent of the scientific community or the federal government, have this responsibility? Should human embryo and cloning research be restricted by the state or local governments, or some combination of these?

Finally, and perhaps more importantly, if there is a market for human embryo and cloning research, and the products of their research, can any type of legislation, at any level, aimed at restricting them be effectively enforced?

A democracy is designed to facilitate a balance between competing interests, to achieve the maximum benefit for the maximum number of its citizens. The introduction of new technology challenges a democratic society to decide who gets what, when, where, and how much. The advent of cloning via nuclear transfer technology presents the inevitability of new and important social changes, and new issues concerning this power, and who controls it, are at hand.

Who, if anyone, should own and control cloning technology? Who, if anyone, should own and control the products of cloning technology?
In the US it is possible to patent both cloning processes and genetically altered, living creatures. In contrast the European Community prohibits the patenting of genetically altered animals, but patenting the process of cloning is possible. Questions concerning the ownership and control of cloned plants and animals, who may not have been genetically altered, have not been answered.

What do genetic engineering, and the cloning of human beings, mean for sociological and legal definitions of, and concepts concerning, the notions of 'individual', 'human', and 'citizenship'?
These will be some of the most difficult and interesting questions that need to be decided. Would a cloned human be an individual? Would it really be a human, with a soul? And what if this clone were then cloned again, and again? What would their status and roles be? Would a non-human primate, such as a chimpanzee, who carried one or more human genes via transgenic technology, be defined as still a chimp, a human, a sub-human, or something else? If we chose to define it as a human, would we then have to give it rights of citizenship? And if humans were to carry non-human, transgenic genes, would that alter our definitions and treatments of them?

Other questions and issues include a revitalisation of the 'nature/ nurture' debate. Will genetically identical people be physically and behaviourally identical, too?
Will cloned humans really look exactly alike? Will they have identical personalities? How will clones impact the future of twin studies meant to ferret out the different impacts of genes vs. the environment? What will human clones be able to contribute to the perspectives of sociobiology?

What about eugenics?
Can we, and should we, use the biotechnologies of genetic manipulation and cloning to improve the human condition? What are some of the perceived risks and benefits of eugenics?

© *National Center for Genome Resources*

Should we clone humans?

Information from the Church of Scotland

Human cloning – will it ever happen?
One of the abiding SciFi nightmares has been the idea that we could one day replicate human beings asexually, just by copying material from human cells. This was one of the most chilling features of Huxley's *Brave New World*. More measured scientific assessments have generally regarded this as something pretty remote. And many in the churches and elsewhere hoped it would stay that way. Roslin's scientists have told a Select Committee of the House of Commons that the nuclear transfer technique they have applied to produce Dolly could be in theory applied to humans. Whether anyone would try and whether it would work is another matter. But the 'what if?' question must now be asked with much more seriousness than would have ever been justified before.

Two aspects of the Roslin discovery have set the world of biotechnology alight. One is the fact that a somatic tissue from an adult has been used to produce a live animal. This has rewritten one of the laws of biology. Up to now it had been assumed that once animal cells go through the mysterious process of differentiation, and become a particular type of cell, they cannot go back to being undifferentiated. Now Dr Wilmut's work has caused a set of cells to forget what they are and start all over again, as if they were undifferentiated. The second is that you can clone a large mammal from the cells of an adult of the species. It is this second aspect that has caught the public imagination, because it has dramatically brought forward the question of whether it could be possible to realise the SciFi dreams of cloned humans.

Faced with such a fertile prospect, the human imagination runs riot, and the media have come up with some very bizarre ideas. One article claimed that we might clone humans to select out genetic defects or select for desirable traits. This would be impossible just by cloning. It might in theory be done by germline gene therapy, but that is quite another, and highly controversial, story. The announcements that nuclear transfer cloning is possible not only in sheep but in cattle and mice suggests that the technique could be quite general in mammals, and thus potentially more likely in humans than when it had been done only on a single sheep.

Scientifically this would be a big and highly dangerous leap to go from cloning a sheep to cloning humans, and it is premature to discuss this as if it were inevitably going to happen. But this discovery means that we have at least got to ask the question, 'What if?'

Why cloning humans is ethically unacceptable

Dr Wilmut, the scientist involved, and his colleagues at Roslin have made it quite clear that they think that to clone humans would be unethical. The Human Fertilisation and Embryology Authority agrees with the general public impression that to clone human beings would be ethically unacceptable as a matter of principle. I and most people in the Church of Scotland would certainly agree that on principle, to replicate any human technologically is something which goes against the basic dignity of the uniqueness of each human being in God's sight. Christians would see this as a violation of the uniqueness of a human life, which God has given to each of us and to no one else. In what sense do we mean this?

Some say that the existence of 'identical' twins means that we should have no ethical difficulty over cloning, or that to object to cloning implies that twins are abnormal. This argument does not hold. Biologically, identical human twins are not the norm, but the unusual manner of their creation does not make them any less human. We recognise that each is a uniquely valuable individual. There are two fundamental differences between cloning and twinning, however. Twinning is a random, unpredictable event, involving the duplicating of a genetic composition which has never existed before and which at that point is unknown. Cloning would choose the genetic composition of some existing person and make another individual with the same genes. It is an intentional, controlled action to produce a specific known end. In terms of ethics, choosing to clone from a known individual, and the unpredictable creation in the womb of twins of unknown genetic nature belong to categories as different as accidental death is to murder. The mere existence of 'identical' twins cannot be cited to justify the practice of cloning.

Controlling someone else's genetic make-up

Thus it is not the genetic identity that is the crucial point but the human act of control, and it is this element of control which provides the fundamental ethical case against human cloning. The biblical picture of humanity implies that we are far more than just our genes, or even our genes plus environmental influences, there is also our spiritual dimension, made in God's image, constituting a holistic notion of being, in which the relational element is as important as the individual. To be a person is to be in relationship. Hence it is vital that the relational implications of technology are considered alongside the ontological. It is against this picture that most Christians would see it as ethically unacceptable to clone human beings as a matter of principle. In so far as genes are a fundamental part of our make-up, to choose to replicate the genetic part of human make-up technologically is a violation of a vital aspect of the basic dignity and uniqueness of each human. By definition, to clone is to exercise unprecedented control over the genetic dimension of another individual. This is quite different from the control parents exert in bringing up their children. Whatever the parents do or do not do, it is inevitable that they have a profound effect on their children. No one exerts the level of control involved in preselecting a child's entire genetic make-up except by a very deliberate act. Moreover, a child can reject any aspect of its upbringing, but it could never reject the genes that were chosen for it. Such control by one human over another is incompatible with the ethical notion of human freedom, in the sense that each individual's genetic identity should be inherently unpredictable and unplanned.

Instrumentality

Cloning raises a number of concerns arising from its consequences, of which instrumentality and risk are of especial importance. To replicate any human being technologically is a fundamentally instrumental act towards two unique individuals – the

one from whom the clone is taken and the clone itself. In nearly all the speculative ideas for cloning a human would use the clone as a means towards someone else's end. They would be created as clones for the primary benefit not of the individuals themselves but of some third party. This would be the case for cloning a dying child or parent to help those bereaved cope with the loss, or cloning an infant with a pre-disposition to leukaemia, as a source of bone marrow which would suffer less tissue rejection problems. These violate a basic ethical principle, that of creating another human being other than primarily for their own sake. There is an important distinction in Christian theology, which admits an instrumental role for animals, to a limited degree, but prohibits it in humans. To clone a child with leukaemia to provide compatible bone marrow would treat the cloned sibling to that extent as means to an end, for the benefit of a third party, rather than for their own sake, and without their consent. Dorothy Werth cited the controversial US case where this was done through normal reproduction, but I would question whether the fact that it worked is justification enough. Again, it is rightly said that we have mixed motives for why we want children, but that does not justify treating a child as a means to an end.

Infertility – an exception to instrumentality

An exception to this objection would be the idea of producing a child from an infertile couple by cloning one of them. But this raises other problems. Instead of being the unique genetic product of both parents, the child is a copy of one of them. For many Christians this would be a denial of a basic relational aspect of re-production, just as in the case of surrogacy. For an infertile couple to have a child by cloning one of them would not normally be thought of as an instrumental act, and might at first sight sound like a compassionate option to offer to childless couples. As observed above, however, there could be serious ethical problems, notwithstanding the anguish which

childlessness brings to many couples. It would not be the biological child of both parents in the normal sense. For many this might be seen as taking the technological harnessing of the desire for a child one step too far, a means which is not justified by the end. The tendency is becoming to demand parenthood as my right, as though it were some moral absolute. We are losing the Christian under-standing of children as a gift, not a right which we can presume that God or life should give us on demand.

Psychological effects – identity and relationship

There are a number of reasons why human cloning might be ruled out for the psychological dangers in-volved. No one knows what would be the effects on human identity and relationships of creating someone who is the twin of their father or mother, but born in a different generation and environment. Would the clone feel that he or she was just a copy of someone else who already existed and not really themselves? Am I really someone else but put into a different womb? What will be my relationship to the one I was cloned from? No one can predict with any degree of assurance what the response would be. Presumably it would vary from person to person. I suggest there are sufficient dangers for applying the precautionary principle. In other words, even though one could not be sure how many people would suffer in this way, it would be wrong knowingly to inflict that risk on someone. Whose interests are being put first?

Physical risk

Dolly took 277 attempts and nearly 30 failed pregnancies to get one success. To repeat the same thing on humans would be giving both the mother and the potential foetus an unacceptably high risk of damage. The basic science of fusing the cytoplasm and nucleus and re-activating the cell is very poorly understood. How many abnormal babies would have to be produced to get one right? There are sufficient unknowns about physical problems

in pregnancy with cloned sheep and cattle to suggest that human cloning experiments would violate normal medical practice. Roslin researchers have said that there is no experiment that could be done to prove the safety of human cloning without causing serious risk to humans in the process. Then there are also unknown factors of ageing. How old is Dolly? Is she her age since her birth, or her age since birth plus the age of the tissue from which she was taken? No one knows the effect of nuclear transfer on ageing processes.

Social risk

Finally, human cloning would bring grave risks of abuses to human dignity and exploitation by unscrupulous people. We have already seen examples of people offering cloning services for large sums of money, when there is currently no reasonable prospect of delivery, and apparently regardless of the risks involved or, in the case of Richard Seed, the rule of law. It is also an open door for abuse, in the way that another individual, a group in society or even the state could exert undue control over an individual. If anyone ever did un-fortunately clone humans, it is import-ant to counter the suggestion from science fiction that they would be sub-human androids with human bodies but no souls. More seriously, some papers from an Islamic perspective seem to imply that if reproduction is by human artifice, it lacks the spiritual element. Some Christians think the same. I do not, however, see any grounds that a cloned child would be any less human than another child. Why would God fail to make the child fully 'in His image' just because of the manner of concep-tion? There would need to be con-siderable safeguards to avoid the risk of stigmatisation. It would be foolish to imagine that abuses could not occur.

Non-reproductive cloning?

The announcement of mouse cloning opens up a whole field of research into other applications of cloning technology which would stop short of full human beings or animals. What applications are envisaged? At present all is speculative, but some of the implications could well be ethically contentious.

Two pieces of research announced in late 1998 suggest that, after many years of trying, scientists appear to have been able to 'redirect' human embryos from their normal course of forming a complete person to becoming only certain types of body cells. Using special cells, known as stem cells, cells of, say, bone marrow or different tissues could be produced more or less indefinitely in the laboratory. In theory, this would open up immense potential for treatments of degenerative diseases where fresh cells could be injected into the patient, in the way that, for example, foetal brain cells have been used on some Parkinson's disease sufferers (currently with uncertain results). One problem with this is rejection by the body, but if the embryo were cloned from cells taken from one's own body, would it be possible to overcome the rejection? No one knows whether any of this can be done successfully and safely.

At the moment it is speculation, but it is seen as important enough to need to know what are the ethical implications of such developments, were they to be possible. During 1998, the UK Human Embryology Authority and Human Genetics Advisory Commission ran a limited public consultation on medical uses of cloning technology, both reproductive and non-reproductive (albeit that these most recent research findings came too late for public comments to be submitted. The joint report *Cloning Issues in Reproduction, Science and Medicine* in December 1998 proposed that the UK regulations should now be amended to allow research into cloning embryos for cell replacement and another non-reproductive application involving mitochondrial disease.

SRT's reaction to this proposal and the ethical dilemmas it poses is given in more detail in *Cloning Human Embryos for Spare Tissues – an Ethical Dilemma*. While we welcome the report's clear conclusion against the reproductive cloning of human beings, we express deep concern about proposals to clone human embryos which would be used

not for reproduction but as a source of replacement tissues. We also call for a much wider public debate of this issue before the UK amends the regulations to allow this controversial new development.

Some have speculated whether it would be possible on the basis of these discoveries to grow not an entire human being, but living organs from cells. Experts at the Roslin Institute and a number of other authorities see this as very unlikely to be possible in most cases. There would be many immense practical questions to answer, and it would be naive to assume that just because we think it, science will one day find a way to do it. This would also raise a number of serious problems ethically. It could be almost as controversial as human cloning. This was brought into focus by the announcement in October 1998 of work in which headless frog embryos were produced by reprogramming cloned cells. This was extrapolated, rather over-sensationally, by the researcher to the possibility of creating separate human organs surrounded by sacks of skin, grown as it were in a test tube from reprogrammed, cloned cells. This was also reported briefly during an excellent BBC TV *Horizon* documentary on cloning. Aside from the technical questions begged by this extrapolation, there is a serious question about whether it would be ethically acceptable to create organs as separate entities from human tissue. This needs careful thought, rather than knee-jerk reactions, but at first sight this would probably be

unacceptable to many people. And this may be only a thought experiment, albeit a dangerous one, since to be of any use, one would have to have such an organ already existing at the same age and stage of development as the patient requiring one. Would this be possible without having a substitute organ already growing from the time of one's birth?

Even if this could be overcome, it could be argued that this was only justified *in extremis* and only for the benefit of the individual involved, or, with appropriate informed consent, a close relative. And underneath this partial cloning lies the ethical problem known as 'gradualism'. By a progression of small steps you could eventually provide all the conditions needed to clone the entire human being, even though that had never been the intention of the research. This raises a much deeper question about how the direction of research is determined and controlled.

What do you do with a genie out of the bottle?

It is impossible to reverse knowledge, but it is society's prerogative to state which pieces of knowledge should remain unused – 'can do' never implies 'must do'. It is rightly illegal to clone a human being in the UK, but it would not be beyond human perversity for someone to try to do it elsewhere in the world. One UK doctor, who has publicly condemned the very idea, claims to have had people offering themselves for

cloning or asking to have it done to loved ones. It would appear that some are attracted by the idea, but maybe they have not fully understood the implications. Even supposing someone would be stupid enough to try, there are many serious obstacles to be overcome. There are firstly the risk of imprisonment. A scientist would take the risk of ostracism from a disapproving medical and scientific establishment, and know that a journal would possibly refuse to publish any paper on the subject. Then he or she would have to persuade or induce dozens of people to take part in prolonged illegal experiments. It would need donors, egg cell recipients and surrogate mothers in fairly large numbers to take part in experiments. Abnormally large progeny have resulted in animal cloning done to date, which suggests that there is serious risks to the health of the mother and any potential embryo.

So are we really supposing all this is almost certain to happen? There is a finite risk, but it is not 100%. According to some, it is well nigh inevitable. Many would not agree. There are many reasons why what is scientifically possible is not always done. It is a commonplace that most active research scientists create far more potential research ideas than they have the time, people and money to pursue. There are presumably many things which it would be scientifically possible to do to live patients that are illegal which would be medically very useful for knowledge of the human disease. This does not mean that they are all inevitably done. Of course it is right to raise the question of cloning of humans as a result of the Roslin research, but let us keep a sense of proportion about the level of risk.

The need for a worldwide ban on cloning human beings

We need to be sure the UK legislation is adequate to outlaw any such activity, about which some have already expressed doubts, and there should be immediate moves to set up, if possible, an internationally binding treaty to ban experiments that would lead to human cloning. The Church of Scotland, along with many others, considers that attempts to clone human beings should be outlawed worldwide. It would be impossible to stop a 'back street' clinic or a dictatorship from ignoring such an international treaty, but the lines need to be drawn. A second line of defence is also called for – the notion of the ethical scientist, for whom it would be against all professional principles to pursue such research. The attitude of the Edinburgh scientists in condemning the idea of human cloning as unethical is a good example.

For more about the legislation and social involvement on cloning, see the SRTP web site page: How do we cope with these questions as a society?

Let's hear it for the animals!

The attention on speculating about cloning humans has also missed something of the point, namely the much more immediate impact of this work on how we use animals. Also see the web pages: Should we Clone Animals? and Cloning, Ethics and Animal Welfare.

• This information was produced by the Society, Religion and Technology Project of the Church of Scotland. See page 43 for details of their web site. ©Donald M. Bruce, 1998

Anti-cloning research

Information from *Conceiving a Clone* – Kayotic Development

To those against it, cloning presents as much a moral problem as a technical problem. For them, cloning is an affront to religious sensibilities; it seems like 'playing God', and interfering with the natural process. There are, of course, more logical objections, regarding susceptibility to disease, expense, and diversity. Others are worried about the abuses of cloning. Cloning appears to be a powerful force that can be exploited to produce horrendous results. Their basic objections to cloning research are outlined here.

Cloning may reduce genetic variability. Producing many clones runs the risk of creating a population that is entirely the same. This population would be susceptible to the same diseases, and one disease could devastate the entire population. One can easily picture humans being wiped out by a single virus, however, less drastic, but more probable events could occur from a lack of genetic diversity. For example, if a large percentage of a nation's cattle are identical clones, a virus, such as a particular strain of mad cow disease, could affect the entire population. The result could be catastrophic food shortages in that nation.

Cloning may cause people to settle for the best existing animals, not allowing for improvement of the species. In this way, cloning could potentially interfere with natural evolution.

Cloning is currently an expensive process. Cloning requires large amounts of money and biological expertise. Ian Wilmut and his associates required 277 tries before producing Dolly. A new cloning technique has recently been developed which is far more reliable. However, even this technique has 2-3% success rate.

There is a risk of disease transfer between transgenic animals and the animal from which the transgenes were derived. If an animal producing drugs in its milk becomes infected by a virus, the animal may transmit the virus to a patient using the drug.

Any research into human cloning would eventually need to be tested on humans. The ability to clone humans may lead to the genetic tailoring of offspring. The heart of the cloning debate is concerned with the genetic manipulation of a human embryo before it begins development. It is conceivable that scientists could alter a baby's genetic code to give the individual a certain colour of eyes or genetic resistance to certain diseases. This is viewed as inappropriate tampering with 'Mother Nature' by many ethicists.

Because clones are derived from an existing adult cell, it has older genes. Will the clone's life expectancy be shorter because of this? Despite this concern, so far, all clones have appeared to be perfectly normal creatures.

A 'genetic screening test' could be used to eliminate zygotes of a particular gender, without requiring a later abortion.

Cloning might be used to create a 'perfect human', or one with above normal strength and sub-normal intelligence, a genetic underclass. Also, if cloning is perfected in humans, there would be no genetic need for men.

Cloning might have a detrimental effect on familial relationships. A child born from an adult DNA cloning of his father could be considered a delayed identical twin of one of his parents. It is unknown as to how a human might react if he or she knew he or she was an exact duplicate of an older individual.

• The above information is an extract from the *Conceiving a clone* web site: library.advanced.org/24355/data/ All research for and development on the web site was done by Kayvon Fatahalian, Bennet Schneider and Brandon Reavis of Kayotic Development. ©*Kayotic Development*

Cloning supporters

Information from *Conceiving a Clone* – Kayotic Development

Supporters of cloning feel that with the careful continuation of research, the technological benefits of cloning clearly outweigh the possible social consequences. In their minds, final products of cloning, like farm animals, and laboratory mice, will not be the most important achievement. The applications of cloning they envision are not nightmarish and inhumane, but will improve the overall quality of science and life. Cloning will help to produce discoveries that will affect the study of genetics, cell development, human growth, and obstetrics. Human cloning is not the issue, it is merely a threat to the continuation of cloning research. Their arguments for such research are displayed here.

Cloning might produce a greater understanding of the cause of miscarriages, which might lead to a treatment to prevent spontaneous abortions. This would help women who can't bring a foetus to term. It might lead to an understanding of the way a morula (mass of cells developed from a blastula) attaches itself to the uterine wall. This might generate new and successful contraceptives.

Cloning experiments may add to the understanding of genetics and lead to the creation of animal organs that can be easily accepted by humans. This would supply limitless organs to those in need. The growth of the human morula is similar to the growth at which cancer cells propagate. If information derived from cloning research allows scientists to stop the division of the human ovum, a technique for terminating cancer may be found.

Cloning could also be used for parents who risk passing a defect to a child. A fertilised ovum could be cloned, and the duplicate tested for disease and disorder. If the clone was free from defects, then others would be as well. The latter could be implanted in the womb.

Damage to the nervous system could be treated through cloning. Damaged adult nerve tissue does not regenerate on its own. However, stem cells might be able to repair the damaged tissue. Because of the large number of cells required, human embryo cloning would be required.

In in-vitro fertilisation, a doctor often implants many fertilised ova into a woman's uterus and counts on

The applications of cloning they envision are not nightmarish and inhumane, but will improve the overall quality of science and life

one resulting in pregnancy. However, some women can only supply one egg. Through cloning, that egg could be divided into eight zygotes for implanting. The chances of pregnancy would be much greater.

Cloning would allow a woman to have one set of identical twins instead of going through two pregnancies. The woman may not want to disrupt her career, or would prefer to have only one pregnancy. With cloning it would be assured that they would be identical.

Cloning could provide spare parts. Fertilised ova could be cloned into several zygotes, one would be implanted and the others would be frozen for future use. In the event the child required a transplant, another zygote could be implanted, matured, and eventually contribute to the transplant. Some believe that if a parent wanted to produce talents in a child similar to his own, cloning using DNA from the cell of the adult may produce a child with the same traits. Many are sceptical about this possibility.

• The above information is an extract from the *Conceiving a clone* web site: library.advanced.org/24355/data/ All research for and development on the web site was done by Kayvon Fatahalian, Bennet Schneider and Brandon Reavis of Kayotic Development ©*Kayotic Development*

Send in the clones?

By Brenton Priestley

In the first few weeks of March, 1997, the headlines were ablaze. *Time Magazine* declared: 'The age of cloning', the *Washington Post* proclaimed: 'British scientists successfully clone adult sheep', *The Australian*'s headlines read: 'Animal cloning sets scientific milestone'. The reason for this media explosion was that scientists had managed to create an identical genetic copy of a mammal, a feat that was previously thought impossible. When the world learnt about Dolly, the cloned sheep, it automatically moved on to the next, and ultimately more important conclusion: if you can clone a sheep, then you probably can clone a person, too.

To understand the issue of cloning, the first thing is to understand what cloning itself actually is. Sci-fi books and movies can create all sorts of wrong perceptions about cloning, with everything from headless monsters to mind control. Cloning is the creation of a genetically identical creature from the DNA of another. In essence, there were clones before Dolly was around, identical twins. Although the DNA of a set of identical twins is the same, and they look very similar, they are still individual separate beings. The difference between a clone and an identical twin is that if I had an identical twin here now, we may be a few minutes apart in age, whereas if I was cloned now, by the time my identical twin was born, we would have a 17-year age difference!

So then, why should we send in the clones? Dr Ian Wilmut, the Scottish scientist who created Dolly, the cloned sheep, suggests that we should clone animals to produce genes and hormones that could help cure diseases such as haemophilia and cystic fibrosis. Harold Varmus, director of the American Institute of Health, suggests that cloning could be used to create more productive livestock and organs suitable for human transplants. And then there comes the big question: Should we clone humans?

While some people are avid sticklers for human cloning, others are equally opposed to it. In the words of the father of cloning, Dr Wilmut, to send in human clones ' . . . would be inhumane'. Soon after it was announced that the sheep had been cloned, and people began to inquire about human cloning, Bill Clinton issued this statement: 'Any discovery that touches upon human creation is not simply a matter of scientific inquiry, it is a matter of morality and spirituality as well . . . Each human life is unique, born of a miracle that reaches beyond laboratory science . . . ' In response, US Senator Thomas Harkin replies: 'To those like President Clinton who say we can't play God, I say OK, fine, you can take your side alongside Pope Paul V who in 1616 tried to stop Galileo. They accused Galileo of trying to play God too.'

Dr Wilmut says that there is no viable reason for cloning humans, but others argue that there are at least two. The first main reason suggested for cloning humans is that we could clone exceptional people. Think of how a clone of Albert Einstein or Isaac Newton would help advance science! Or think about sport and movies for the next 50 years if we cloned Michael Jordan or Steven Spielberg!

The other, perhaps more important reason for cloning humans would be to allow childless couples to have genetically related children. Dr Richard Seed, an American infertility expert, made headlines late last year, when he announced that he was going to open a cloning clinic for this purpose. His reasoning behind it was this: ' . . . God made man in his own image. God intended for man to become one with God. Cloning and the reprogramming of DNA is the first serious step in becoming one with God.'

While Seed considers that God would support human cloning, many religious leaders have spoken against it. The Vatican issued a statement specifically condemning the cloning of humans. In fact, religion is one of the main reasons why people object

to human cloning. As well as this there are many other reasons. Some people, like Bill Clinton, consider cloning to go against the sanctity of human life. Some think that as well as gifted people being cloned, people like Adolf Hitler or Saddam Hussein could also be, and in fact, Mr Hussein unveiled his plans that he wanted to be cloned. Some think that people could create monsters through cloning, or that cloned children could turn out to be retarded or deformed. Some think that people could clone themselves solely for the purpose of obtaining body parts.

Some think that if couples want a baby, why not adopt one rather than go through the risky process of cloning? And one of the most common objections to human cloning is a queasy, uncertain feeling, best summed up in the words of Australian scientist, Paul Davies: 'Most people have a natural repugnance for genetic manipulation, especially when it comes to plans for human cloning, yet they find it hard to pin down exactly what is wrong with it.'

The founding father of cloning, Dr Wilmut, is certainly against human cloning. In Adelaide recently for the 14th Australasian Biotechnology Conference, I went to hear him speak. His message was loud and clear: society, ethics, morality and science would not tolerate human cloning. While he said that it was appropriate to clone animals, he and his colleagues could not think of a reason to clone humans. Over the next 20 years, as a result of animal cloning research, he hoped to see cures to such afflictions as Parkinson's disease, muscular dystrophy and nerve damage.

The topic of cloning is certainly a controversial one, and at the moment many people are grappling with it. The best solution to the problem now is to neither totally ban cloning nor to totally encourage it, but discuss it, research it, and wait a little longer before we decide whether or not we should send in the clones.

• Brenton Priestley is a student from Australia. He had to present a 7-minute speech that discussed the topic 'Send in the Clones'. Although he is pro-human cloning, he had to present both sides of the issue. He says that 'The research that I did for this speech was invaluable in strengthening my support for human cloning, and your site was invaluable in my research.' The above is an extract from the web site www.humancloning.org

© Brenton Priestley

Ethical issues in cloning humans

Kate Cygnar from
Minneapolis

There are three main areas involving the ethical issues of cloning humans, and they resemble ethical questions already in existence. The possibility of cloning has not created new ethical issues, but revived old ones.

Many people talk about the possibility of using a cloned zygote from a parent and implanting it in a host mother, as fertilised eggs are during in-vitro fertilisation. This raises the same issues surrounding in-vitro fertilisation and fertility drugs that were first raised in the media when the 'Iowa Septuplets' were born. In-vitro fertilisation and fertility drugs have the possibility of creating abnormal multiple pregnancies. It is dangerous to the mother and the foetuses to carry a pregnancy where multiple foetuses are involved to term. Many people involved in multiple pregnancies often choose the selective reduction procedure, where some of the foetuses are removed. This practice also carries a wide range of ethical issues and is difficult for the mother emotionally.

Another issue raised by the possibility of human cloning is that of social Darwinism, a belief that less desirable human beings will be eliminated (the Nazi party used this belief to rationalise the extermination of Jewish people during the Holocaust). Some individuals have expressed the idea of cloning individuals with certain talents (i.e. Wolfgang Amadeus Mozart or Albert Einstein) to 'better' society. Most people find this very offensive, which brings us to my third topic: our prejudice against those who are different.

How would it feel to know you were a clone? I don't think anyone can argue that society would not hold 'clonist' views against you, treating you as if you were a different type of person. What would be worse, perhaps, would be to be a person cloned with a specific person in mind for you to imitate. It is as likely as it is not that a clone of Michael Jordan would hate basketball, but if society defined his purpose in life to be a star basketball player he would have a very miserable life. What if Michael Jordan II had no talent for basketball? It is entirely possible that such a trait as talent for basketball is something that developed randomly during development. Yet people would expect him to have a talent for basketball, and would hold him to standards which they would not dream of holding to another child. Also along this point – what if everyone in basketball was Michael Jordan? Would basketball be as intriguing? Would any non-Michael Jordan ever have any interest or even chance to play at the sport?

I ask you to consider these points. As of yet there has been no demonstrably 'good' use for the cloning of humans, so I say, hold off on human cloning. We have better things to work on.

Cloning and eugenics

I briefly touched upon this above, and since I have been receiving many letters saying that cloning would be okay if we cloned someone like Isaac Newton, I would like to reiterate my point. In order to understand my purpose, one must first understand what eugenics is all about.

The eugenics movement began just after Charles Darwin proposed his theory of evolution by natural selection. A man by the name of Francis Galton began the movement – his goal was to accelerate the natural selection which must be occurring against the undesirable members of human society. Until the Americans picked up on Galton's ideas around the turn of the century, the movement was relatively harmless, only encouraging upper-class families to have many children.

Also at the turn of the century, the papers of Gregor Mendel, the Austrian monk who did extensive work on how traits were passed on in generations of peas, came to light. Now that a mechanism for heredity was proposed, the eugenics movement took off. The movement was based on what is now considered extremely faulty logic. Things such as 'feeblemindedness', poverty, lower social status, low intellectual ability, and blindness were all considered to be hereditary. Different races were also supposed to have different intellectual capacity, such as the Poles, Hungarians, and blacks. (This lines up with a large influx of immigrants from eastern and southern Europe into America.) The theories of eugenics were used to support the anti-immigrant sentiment that Americans feel during a wave of new immigration, such as the Irish in the mid-1800s and the Latinos and Asians today. The eugenics movement led to many anti-immigration laws as well as many marriage restriction acts and sterilisation acts, restricting the marriages of people deemed unfit to have children (because of 'feeble-mindedness', other mental disorders, physical handicaps, etc.) and the sterilisation of such people in mental institutions. The Germans beat the Americans at their own game, with a movement which eventually turned into the genocide of the Holocaust.

Most people agree the eugenics movement of the early 20th century was a bad thing. The ideas expressed when someone says that we should reproduce a person genetically identical to Newton are very similar to the ideas of the eugenics movement.

(1) You assume that the traits of such a person were genetic, not a result of random developments during a person's growth from unicellular to multi-cellular (such as fingerprints) and not a result of other non-genetic traits such as motivation or environmentally determined traits.

(2) You expect that society will be improved by having more people like that in it.

• The above information is an extract from the web sites cloning.cjb.net/ and members.tripod.com/~kateDC/clone.html

Police warn of illegal cloning

An illegal trade in cloned body parts and genetically engineered children has been identified by an elite British police squad as a future market for organised crime.

The head of a new research unit at the National Criminal Intelligence Service (NCIS) also issued a warning yesterday about money laundering and illegal gambling on the Internet as well as a potential new addiction to 'virtual reality drugs'.

Crime groups are already becoming involved in the international sale of transplant organs, such as kidneys, said Robert Hall, the head of analysis at NCIS. 'Genetic commerce also has a very lucrative potential not only for the scientists but also the unscrupulous practitioner or criminal who wishes to make easy money. At its simplest,

**By Jason Bennetto,
Crime Correspondent**

there is the organised criminal who sells illegally acquired, genetically engineered body parts.

'At the more complex, there is the bogus agent offering gene therapy to the unwary or desperate parent, or even customising genetic changes into a newly conceived child.'

> **'By 2020, 95 per cent of human body parts could be replaceable by laboratory-grown organs'**

Speaking yesterday at a conference on international crime held in London, Mr Hall described some of the developments that could be illegally traded in the future. 'By 2020, 95 per cent of human body parts could be replaceable by laboratory-grown organs. Cloning will enable parents to select and improve the character and health of their children.'

The NCIS is also investigating organised crime's role in environmental issues, such as dumping toxic waste, and the trade in endangered species.

'There is evidence that toxic dumping is very prevalent in the United States. We know it happens here, but we do not yet know the extent of the problem,' Mr Hall said.

Cloning: is this the future for farm animals?

Information from Compassion in World Farming (CIWF)

What is cloning?

Cloning is the production of a precise genetic copy of an animal, human or plant, or of some part of them. The genetic material of an animal is contained in the nucleus of every cell in its body. The well-known cloning experiments on sheep at the Roslin Institute in Scotland in 1996 and 1997 all involved methods where the genetic material is removed from a cell of one animal and put into the egg-cell of another animal. This is called 'cloning by nuclear transfer'.

As in the case of the lamb Polly, born in 1997, the genetic material can also be genetically engineered before it is transferred to the egg-cell. Polly was genetically engineered to produce a human protein in her milk. The scientists saw this as potentially a way to produce an 'instant flock' of transgenic (genetically engineered) animals.

Since the birth of Dolly the pace of cloning experiments on sheep and cows all around the world has been speeding up. The commercial aims are twofold: one is to produce human proteins more cheaply in the milk of transgenic animals for pharmaceutical uses (known as 'pharming'). The other is a method of 'livestock breeding'. As we go to press, the latest cloning announcement in December 1998 was that 8 calves had been cloned from the cells of a beef cow at Kinke University in Japan. Four of the calves died within 3 days of birth. The scientists said that they hope to be able to duplicate cows 'proven to be ideal milk and meat producers'.

CIWF has always believed that the cloning of farm animals is completely incompatible with today's understanding of animals as individual sentient beings, capable of suffering.

What happens to animals in cloning experiments

Let's consider how Dolly was born. The cell containing her DNA belonged to a 6-year-old ewe who had been killed some years before. The ewes who provided the egg-cells were given hormone injections to stimulate superovulation and their eggs were removed by surgery. The 'reconstructed' embryos were put temporarily into the oviducts of other ewes to start their development. The ewes were then killed and the embryos taken out and checked. A total of 156 checked embryos were transferred to a total of 64 'surrogate mother' ewes. Very few of these embryos survived. At day 110, 4 foetuses were found to be dead. The ewes were killed and 2 of the foetuses were found to have abnormal livers. Finally 8 lambs were born, 1 died, and Dolly was the only one who was a clone from an adult cell.

Abnormal births

In the experiment that produced Polly, only one of the 11 surrogate mother ewes who actually gave birth had a normal delivery. Four had caesarian sections and 6 were induced because of prolonged gestation. Eight of the 14 lambs born died within 2 weeks.

Another example comes from Colorado State University. The scientists reported in 1996 that they had produced 40 cloned calves, many with abnormal birth weights. The heaviest (67.3 kg) could not stand without support. Six others had limb contraction problems. Thirty-four needed medical treatment and 8 died.

Why CIWF opposes the cloning of farm animals

Cloning experiments are just that – experiments. No one knows in advance whether they will work or just what effect they will have on the animals. The experiments consistently tend to produce oversized or abnormal foetuses, a high death-rate of foetuses and a low 'success rate' in terms of normal offspring. The Farm Animal Welfare Council's recent report considers these problems so serious that it has called for a moratorium on cloning in commercial agriculture until they have been solved. The experiments involve subjecting many animals to non-therapeutic invasive procedures and surgery and the killing of egg-cell 'donors' and temporary mothers. CIWF believes it is wrong to subject farm animals to the pain and stress of procedures that are not for their own benefit.

Even if the 'efficiency' of cloning methods improves, cloning technology is still likely to be a welfare disaster for farm animals. It is the latest in a series of technologies designed to manipulate farm animals for our convenience. Selective breeding already has a very bad track record for the health and welfare of farm animals. The creation of herds or flocks of cloned animals could also lead to an even greater loss of genetic diversity with unforseeable results in terms of increased illness for farm animals. A herd of cloned animals genetically engineered to have resistance to one disease could turn out to be very susceptible to another one.

CIWF believes that the suffering involved in the cloning of farm animals cannot be justified by the benefits claimed by the scientists and the multinational companies which back them. We believe that the scientific community should stand back from exercising their cloning technology on sentient animals and that governments should stop funding it now and regulate it out of existence.

© Compassion In World Farming

Cloning – how should society decide?

What do you do with a genie out of the bottle?

It is impossible to reverse knowledge, but it is society's prerogative to state which pieces of knowledge should remain unused – 'can do' never implies 'must do'. It is rightly illegal to clone a human being in the UK, but it would not be beyond human perversity for someone to try to do it elsewhere in the world. One UK doctor, who has publicly condemned the very idea, claims to have had people offering themselves for cloning or asking to have it done to loved ones. It would appear that some are attracted by the idea, but maybe they have not fully understood the implications. Even supposing someone would be stupid enough to try, there are many serious obstacles to be overcome. There is firstly the risk of imprisonment. A scientist would take the risk of ostracism from a disapproving medical and scientific establishment, and know that a journal would possibly refuse to publish any paper on the subject. Then he or she would have to persuade or induce dozens of people to take part in prolonged illegal experiments. It would need donors, egg cell recipients and surrogate mothers in fairly large numbers, to take part in experiments. Abnormally large progeny have resulted in animal cloning done to date, which suggests that there are serious risks to the health of the

mother and any potential embryo.

So are we really supposing all this is almost certain to happen? There is a finite risk, but it is not 100%. According to some, it is well nigh inevitable. Many would not agree. There are many reasons why what is scientifically possible is not always done. It is a commonplace that most active research scientists create far more potential research ideas than they have the time, people and money to pursue. There are presumably many things which it would be scientifically possible to do to live patients that are illegal which would be medically very useful for knowledge of human disease. This does not mean that they are all inevitably done. Of course it is right to raise the question of cloning of humans as a result of the Roslin research, but let us keep a sense of proportion about the level of risk.

Comment on the Korean claim – need for international research guidelines

Research in this area is proceeding at a rapid rate. Developments in stem cell and cloning technology are changed the picture of the issues under debate. This is brought home further by the claim on December 15 that Korean scientists have cloned a very early human embryo. At the moment we should be sceptical about this claim. It has not been published in a scientific journal and the Roslin

Institute says that since the experiment only went as far as the 4-cell stage, it had not reached a stage where it was possible to say that it was a cloned embryo. Nonetheless, the fact that they have apparently tried raises exactly the ethical questions of whether we should allow the creation of cloned embryos which we have posed above.

The need for legislation – a worldwide ban on cloning human beings

The faster science and technology proceed, the less our frameworks of legislation and regulation seem to cope with them. As soon as Dolly was announced already people were asking whether even our present UK Human Fertilisation and Embryology Act covers the possibility of human cloning from adult cells. There is clearly great diversity over this matter – some countries' laws do indeed outlaw human cloning, but in many no effective legislation exists. In the US, only publicly funded research in this area is banned. There has been much progress as many governments have examined the adequacy of their legislative and regulatory situation.

The Church of Scotland, along with many others, considers that attempts to clone human beings should be outlawed worldwide. There seems to be a substantial political mandate for this. It would be impossible to stop a 'back street' clinic

or a dictatorship from ignoring such an international treaty, but the lines need to be drawn. The situation on non-reproductive uses and research is much more complex, but it would be important at the very least to work towards establishing some internationally agreed guidelines on what should and should not be allowed in such research. A number of bodies could address this question. The SRT Director was part of the UK delegation at a summit meeting of national bioethics committees in November in Japan. A committee was set up to see what collaborative work could be done. Here is an excellent subject which it could address. Dr Bruce was also recently invited as an observer to the meeting of the International Bioethics Committee of UNESCO. In 1997 this committee produced a Universal Declaration on the Human Genome, together with a statement against cloning human beings, recently endorsed by the UN General Assembly. Again, this should now address the question of cloning research. The Council of Europe's Bioethics Convention, which also has a Protocol on Human Cloning could do the same.

Consideration also needs to be given about how much further research of this type should be done in the animal field. It would seem unfair not to allow the next step of the Roslin work, which would try to attempt the same exercise on genetically engineered cells, but how much further beyond that should it go? Unfortunately there is a lack of necessary procedures to look at the ethics of animal experiments, as opposed to safety or welfare. The existing UK animal experimentation legislation is not designed to allow for the ethical assessment of experimental proposals. Moreover, the continuing lack of a statutory Ethical Commission on Biotechnology is leaving such issues to be dealt with on a piecemeal basis, when many are asking for a better way of balanced debate.

Ethical scientists

But is legislation the answer, when someone could in theory find a way of getting round the laws or simply

doing illegal research in secret? How is the research itself controlled? It would be unfair to imply that Roslin scientists have been unnecessarily hiding things away. A second line of defence is also called for – the notion of the ethical scientist, for whom it would be against all professional principles to pursue such research. The attitude of many scientists in condemning the idea of human cloning as unethical is a good example, but it needs to go beyond this. There needs to be a general drive aimed at scientists, both those practising today and those being trained, developing a more thorough appreciation of the ethical and social dimension of their work. Gone are the days when we could agree to the notion of 'disinterested' science, in which the researcher went into the lab and left his or her values outside the door, in order to be 'objective'. Scientists' values inevitably come into their work, but few are given much guidance in thinking ethically about their work. It should be a part of any science degree course and especially a requirement of the membership of professional scientific, engineering and medical institutions for scientists to have done courses in ethics and to demonstrate in their work their credentials as ethical scientists. Ethics should be just as much part of the scientist's bag of tools as the ability to draw a graph.

Openness of research

The third line of defence is openness to public scrutiny of the research and its aims. The pace of present developments raises the point of whether society has an effective scope for comment on the acceptability of research which would break new ground ethically. For some time it has been clear that we need a better way of enabling the public assessment of new areas like this where there could be cause for concern. Ultimately society as a whole must own the research which is nominally being done on its behalf. This can only happen if there is true participation. Ethics committees are useful up to a point, but they can never be the complete answer for how to involve the public and provide for a debate which engages the non-expert. There needs to be a radical improvement of providing for informed public debate on developments of this nature.

For further information

This information has been produced by the Society, Religion and Technology Project of the Church of Scotland.

We'd welcome any comments you may have. We don't claim to have said the last word!

If you want to send us a comment or obtain further information or receive our latest Newsletter, email us at : mailto:srtscot@dial.pipex.com

• This information has been produced by the Society, Religion and Technology Project of the Church of Scotland. See page 41 for address details.

Should we clone animals?

Information from the Society, Religion and Technology Project (SRT) of the Church of Scotland

Since 1993, a working group of the Society, Religion and Technology Project (SRT) of the Church of Scotland has been examining ethical issues of genetic engineering. This inter-disciplinary group of experts includes Roslin scientist Dr Ian Wilmut, head of the team who cloned Dolly. The following article by the Director of the SRT Project reflects on some of the ethical issues, straight from the sheep's mouth.

Dolly and her applications

Dolly is the most famous sheep in the world. She looks much like any other sheep, but she has been cloned from another adult sheep. Scientists at the Roslin Institute in Edinburgh have rewritten the laws of biology. Her announcement in February 1997 led to an unprecedented media circus which caused as much confusion as it shed light. The attention focused mainly on speculations about the possibility, or otherwise, of cloning humans. In doing so, it missed the much more immediate impact of this work on how we use animals. It's by no means certain this would really lead to flocks of cloned lambs in the fields and hills of Scotland, or clinically reproducible cuts of meat on the supermarket shelves. But it does prompt us to ask questions about the way we are using animals with new technology, and the kinds of assumptions we make.

Cloning had already been done to a limited degree by splitting embryos, mostly in cattle, and raised ethical and welfare concerns in the process. But the Roslin work opens up the prospect of a far wider range of applications from adult animal cells. At the moment, there are only a few early results in sheep, and rather little is understood of how it has happened. Different farm animal species differ somewhat in their embryology. Now the technique has been extended to cattle and also mice, suggesting that it could be general in mammals. It remains to be seen to what extent the method would work in different animals without adverse effects. But assuming it could be applied more widely, what are the potential applications in animals?

Since 1986, Roslin have been genetically modifying sheep to produce proteins of therapeutic value in their milk. Successful as this has been so far, the present methods are very hit and miss, using perhaps 100 live animals to get just one right one. The original aim of Dr Wilmut's nuclear transfer work was to find more precise methods by genetic modification, via a cell culture, if a way could be found to grow live animals from the modified cells. Their announcement in July 1997 of the transgenic cloned sheep Polly marked the first evidence of this principle. The fact it was a clone was, in a way, a side-effect. PPL Therapeutics, the Edinburgh firm behind the research, say they might clone 5-10 animals like this from a single genetically modified cell line, but then breed them naturally, as 'founders' of a set of lines of genetically modified animals. There would be no advantage in cloning beyond the first point.

But these medical applications on farm animals tend to be small-scale affairs. The amount of animals and the amount of milk are very small compared with conventional meat or bulk milk production. Imagine you are a commercial breeder of cows or pigs, and over many generations you have bred some fine and valuable beasts with highly desirable characteristics. One possible application of Roslin's work could be to clone such animals from the cells of one of them, and sell the cloned animals to 'finishers' – those farmers who simply feed up the animals for slaughter, rather than breed them to produce more stock. Again, the breeder might want to clone a series of promising animals in a breeding programme, in order to test how the same 'genotype' responded to different environmental changes. The extension of cloning to mice opens up potentially large areas of research, because mice are so much easier to use, to breed and their genetics and metabolism are better understood.

Ethics and animal cloning

Should this be allowed ethically? To look at this, here are several possible criteria – unnaturalness, diversity, fundamental concerns, animal welfare and commodification.

Is it unnatural?

Many people say that cloning farm animals would be unnatural. Whereas in the plant kingdom cloning is a fairly common phenomenon, there are few animal examples and none in mammals or humans. Should we then respect this biological distinction, or should we celebrate our human capacity to override such limitations? It is hard to argue in an absolute sense that anything is unnatural, when so little remains around us that we might justifiably call natural, and nature itself is in constant motion. Yet many believe some technological inventions are now going too far to remain in tune with what we perceive 'natural' to mean, despite how much we have intervened in nature to date. Is cloning animals a point to draw a line?

Would it narrow genetic diversity too far?

This brings us to the question of diversity. One of the fundamental rules of selective breeding is that you must maintain a high enough level of genetic variation. The more you narrow down the genetic 'pool' to a limited number of lines of, say, animals for meat or milk production, the more you run risks of problems from in-breeding. If that is the case with breeding, how much more is it true of cloning, where genetic replicas are involved? This means there are pragmatic limits to how useful cloning would be, but beneath the pragmatics there lies a deeper ethical concern. Does this reflect something fundamental about the nature of things?

Is there a fundamental ethical concern?

This is something for which Christian theology provides some insights. For the Christian, the world around us is God's creation, and one of its most characteristic features is variety. The biblical writers make repeated allusions to it, painting striking pictures of a creation whose very diversity is a cause of praise to its creator. It could be argued that to produce replica humans or animals on demand would be to go against something basic and God-given

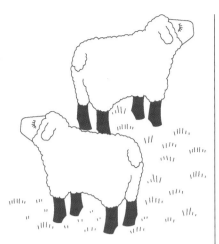

about the very nature of higher forms of life. Where God evolves a system of boundless possibilities which works by diversification, is it typically human to select out certain functions we think are the best, and replicate them? Deliberate cloning aims at predictability, replication, in order to exercise control, whose centralised, even totalitarian approach contrasts with God's command to animals and humans to 'be fruitful and multiply'. In the limit this argument would mean that cloning would be absolutely wrong, no matter what it was being used for. This intuition runs deep in many people. But there are also questions of scale and intention to consider.

Justifiable uses of cloning?

Cloning animals might be acceptable in the limited context of research or, where the main intention was not the clone as such but growing an animal of a known genetic composition, where natural methods would not work. Roslin's work to produce Polly the transgenic cloned sheep would be such a case, where the intention is not primarily to clone, but to find more precise ways of animal genetic engineering. Indeed, producing medically useful proteins in sheep's milk is one of the least contentious genetic modifications in animals, since the intervention in the animal is very small for a considerable human benefit. Careful scrutiny would be needed, to see that it was only applied to genetic manipulations that would be ethically acceptable, but that is a question we already faced before cloning.

Animal welfare concerns

We also need to be sure about the animal welfare aspects even of limited cloning. Questions have been raised about the number of failed pregnancies and unusually large progeny which appear to be resulting from Roslin's nuclear transfer experiments to date. While the suffering is not so great as to put a stop to this work, it is clearly necessary to understand the causes and establish whether the problems can be prevented, before the methods could be allowed for more general use. If after a reasonable time there seemed little prospect of doing so, however, one would doubt whether it was ethical to go any further. This also points to the serious possibility that any attempt at human cloning could be extremely dangerous for both the clone and the mother, and thus medically unethical, irrespective of wider ethical concerns.

The extension of cloning to mice means that many more animals are likely to be used in research at a time when the trend is to reduce animal use. There is a difficult tension here.

Are there unjustifiable uses of animal cloning?

If an ethical case can be made in favour of Roslin's limited and indirect application of cloning, it is a different matter to apply cloning directly in routine animal production, to accelerate or side-step natural methods. For many, this would be unjustifiable, quite apart from the welfare concerns. What's the problem, you might ask, since we already intervene in nature in selective breeding, and use methods like artificial insemination and embryo transfer? If there was a clear benefit to the farmer to start off with prime stock, to produce the best beef or pork, this might seem to have its attractions. But the answer might lie in a wider question about where we have reached in our human use of animals.

Animals as supermarket commodities?

What should we do with animals? Most of us eat them, but not everyone. Quite of a lot of us enjoy them as pets and companions, or watching some of them in the wild. We used to use them to carry and haul for us, until technology made it redundant. But technology is now coming up with other ways of using the creatures we share the planet with, which pose important questions. And whatever use we find for animals, should we clone them so we can do so more efficiently?

One assumption is that the animal kingdom is there for us to use in almost any way scientists dream up or commercial companies see a market, short of inflicting gratuitous pain. The fact that we kill animals to eat them is taken to justify more or less any other use, especially if we can cite human medicine or job creation as goals. On this view, only if they are warm and furry, or primates, do we start to have some qualms, and even then, very selectively.

Many people would disagree. Nature is not ours to do exactly what we like with. On a Christian understanding, all creation owes its existence ultimately to God. This does not mean that we cannot use animals, but it does mean that humans have a duty of care and respect towards them, as creatures which exist firstly for God, and which only secondarily may be used by us. Such use must be responsible and with a dignity due to another of God's creatures, and we should hold back from some uses. Is cloning then the point to say 'no'?

The suggestion that cloning is justified because we already intervene so much in animals can be an excuse for looking properly at the case in point. It also begs the question about what we are already doing. There are a number of techniques in regular use on farm animals which are ethically borderline, which illustrates a general problem. Both biotechnology and industrial production methods in agriculture carry over certain assumptions from the sphere of chemistry or engineering which, though scientifically applicable to animals, may not always be morally applicable to them. We see this in the animal welfare problems which conventional selective breeding has caused in some cases, such as poultry, from applying production logic too far.

Against that context, if anything, what is called for is greater restraint. Why would we want to clone meat-producing animals, anyway? Most of the suggested applications relate to production improvements rather than clear human or animal benefits. To create genetic replica animals routinely, for the sake of production convenience for the supermarket, would be to apply a model derived from factory mass production too far into the realm of living creatures. In the limit, to manipulate animals to be born, grow and reach maturity for sale and slaughter at exactly the time we want them, to suit production schedules suggests one step too far in turning animals into mere commodities.

• The above information is from the Society, Religion and Technology Project of the Church of Scotland.
© Donald M. Bruce, 1998

$5m for lab to clone pet dog

Cloning scientists in Britain are receiving requests from owners of dogs and cats wanting copies of their pets.

News of the attempted deals emerged yesterday when it was disclosed that a millionaire in the United States has paid $5 million to clone a mongrel called Missy.

According to Mark Westhusin, director of the cloning laboratory at Texas A & M university, scientists have been given two years to duplicate the dog from a few cells. Mr Westhusin told BBC2 *Newsnight* last night he thought there were other millionaires who would do the same. 'I'm sure there are lots of them out there that would, if they knew the potential existed.' He also confirmed that his laboratory had been approached to clone racehorses as well as pet rabbits.

In May, a firm called Clonaid – connected to the Raelian religious movement which has an extra-terrestrial theme park called Ufoland

---
By Tim Radford,
Science Editor
---

in Quebec – announced a service called Clonapet.

But right from the start the Roslin Institute in Scotland, which last year made history by cloning Dolly the sheep, has tried to stop people believing that cloning is a form of resurrection. A clone would have the same genes – but a different womb, a different environment, and different life history. The result would be a different identity.

A millionaire
in the United States
has paid $5 million
to clone a mongrel
called Missy

Harry Griffin, of the Roslin Institute, said: 'We have not been offered large sums of money, but we certainly have had approaches by people who want to clone their pets.'

Cloning began more than a decade ago – but Dolly made history because she was cloned from an adult cell taken from an already-dead ewe. The news immediately raised the spectre of clones made from Adolf Hitler or Shergar the vanished racehorse. One doctor claimed a woman had asked if she could have her father again as a baby. But cloning is still a hit-and-miss affair: it took 277 attempts and 27 pregnancies to produce one Dolly.

'This sort of proposal perpetuates myths about cloning,' said Dr Griffin. 'An individual is a lot more than its genes. The idea that you can recreate your pet is false.'
© *The Guardian*
August, 1998

First male mouse clone may save endangered species

By Catherine Lazaroff

Using cells from the tip of a mouse's tail, scientists at the University of Hawaii at Manoa have produced the first male mice clones, increasing the possibility that the technique could someday aid conservationists in preserving nearly extinct species.

Researchers Teruhiko Wakayama and Ryuzo Yanagimachi say their work, which is reported in the June issue of the journal *Nature Genetics*, demonstrates that mammals can be cloned from non-reproductive cells. Previous mammalian clones, including the famous sheep named Dolly, were produced using cells associated with the female reproductive system.

This research marks the first time that a peer-reviewed study has conclusively shown that it is possible to clone a male mammal.

The researchers used the Honolulu Technique of replacing the genetic information in an egg cell with the nucleus of a donor cell from another individual. Last year, the same team reported producing several cloned female mice with the technique by using cumulus cells, the cells that surround developing eggs within ovaries, as donor cells. Yet some experts asked whether the success could be credited to the animals' sex, and the choice of donor cells.

In an interview with ENS, Professor Yanagimachi said that many people asked his team whether they could repeat their success with a male animal by using somatic cells, the non-reproductive cells that form most of the body.

Yanagimachi also had a particular interest in male cloning because the U. of Hawaii hosts a bioengineering programme involved in researching transgenic animals, genetically engineered animals that produce medical and pharmaceutical products within their bodies.

'We have a transgenic animal that is the only individual that remains of its particular line,' Yanagimachi said. 'It is a male mouse that has a special gene. The mouse is infertile, so we can't get sperm out to reproduce him. Our only hope to save him is to clone him.'

> **This research marks the first time that a peer-reviewed study has conclusively shown that it is possible to clone a male mammal**

By demonstrating that a male mouse can be cloned from cells in the tail, the team brought the hope of saving that transgenic mouse one step closer to fruition. Yanagimachi also believes their technique might someday be used to clone sterile individuals from endangered species, preserving their genes to increase the diversity of the remaining gene pool.

But there are still many obstacles to overcome. Yanagimachi's transgenic mouse is very old, and 'cloning seems to be more difficult as age advances,' he says.

Also, the success rate for the Honolulu Technique remains very low. The team only produced three live births from hundreds of attempts, and two of those died shortly after birth. The single remaining clone has survived to sire two litters of offspring.

'People are working hard to improve the techniques,' Yanagimachi says. 'We are now investigating what's gone wrong with aborted foetuses.

Questions also remain about the environmental effects of genetic engineering. Recent reports that pollen from a bioengineered corn could kill monarch butterflies reignited the storm of criticism that followed the birth of Dolly and many biotechnology advances since then.

Yanagimachi acknowledges that there is 'a possibility' that bioengineered mammals could somehow be detrimental to the environment. 'There are many, many unanswered questions,' he says. 'But most mutants are not strong,' and therefore would be unlikely to survive outside the laboratory to cause harm.

'Of course, we must be careful of the consequences,' cautions Yanagimachi. But he is hopeful that new advances will make cloning more practical and useful in the future. 'We need another breakthrough,' he says. 'We pass one barrier, and there is another one waiting for us.'

Genetic flaws hit cloned animals

Cloned animals have been found to suffer from serious genetic defects – a discovery that could deliver a fatal blow to hopes of ever using cloning for human reproduction.

French scientists have for the first time found unequivocal evidence that cloning interferes with the normal function of genes, a factor that can lead to debilitating illnesses and death. Ian Wilmut, the scientist at the Roslin Institute near Edinburgh who cloned Dolly the sheep, said the findings are the most detailed so far to explain the side-effects resulting from the cloning process.

Professor Wilmut, who cloned Dolly by transferring the nucleus of an udder cell taken from a six-year-old ewe into an unfertilised egg that had its own nucleus removed, said inherent problems with the technique may prevent it ever being applied to humans.

'It astonishes me that people would consider human reproductive cloning and this research adds further concern,' he said.

'It is the most detailed information to emerge so far of the abnormalities arising from nuclear transfer and it is further evidence that we should be extremely cautious in ever applying this to humans.'

Dolly, the first adult clone of a mammal, raised the prospect of scientists being able to clone identical copies of adult human beings. Although Dolly herself appears normal, evidence that other animals created by the same process suffer from genetic disorders will make human reproductive cloning less likely to receive ethical consent.

The French team, led by Jean-Paul Renard of the Institut National de la Recherche Agronomique in Jouy-en-Josas, investigated the death of a calf that was cloned from a skin cell taken from the ear of a 15-day-old calf, which was itself a clone of a bovine embryo.

*By Steve Connor,
Science Editor*

The calf that died had developed normally for the first six weeks but then suffered a rapid depletion of blood cells and severe anaemia caused by the incomplete development of its lymph glands.

> *Scientists have for the first time found unequivocal evidence that cloning interferes with the normal function of genes, a factor that can lead to debilitating illnesses and death*

In a research paper published in the *Lancet*, the scientists say: 'This is the first report of a long-lasting defect associated with somatic (adult cell) cloning.'

Because the 'parent' of the calf was itself an embryonic clone that had suffered no ill effects, the scientists were able to conclude that the death of the calf must have been due to the process of nuclear transfer from an adult cell.

Professor Wilmut said the cause of the problems could be connected with the genetic 'reprogramming' of the adult cell nucleus, which is necessary for it to switch on all the genes needed to create a fully grown animal from a single cell.

'What has to happen is that the adult cell's genes are switched off and the genetic reprogramming needs to be done.

'People were surprised this could happen at all, so it is not surprising that sometimes it almost happens but not quite enough,' Professor Wilmut said.

Cloning by nuclear transfer is known to cause an unusually high number of deaths. Up to 50 per cent of cloned sheep foetuses die in the womb – 10 times higher than normal – and about 20 per cent of live births result in the animals dying within the first few days, about three times higher than the normal death rate immediately after birth.

© *The Independent*
April, 1999

Opinions on animal cloning

Information from the Science & Technology project at the University of Virginia

The cloning of multi-celled animals is now a reality. Questions must now be asked as to how this technology is to be applied. There are many opinions on this issue being surfaced by the various constituencies involved in animal cloning. On one side of the issue are the research and development companies that are doing the cloning, and on the other side are religious concerns worried about the moral and ethical implications involved in the cloning of animals.

Each of the companies that has conducted extensive research in the field of nuclear transfer has goals for how cloning technology should be applied. The possible benefits of their work are not yet fully realised, however several have been suggested. For example large animal breeders have expressed interest in this technology in order to use clones to immediately improve their stock of animals for food purposes. The idea is that clones could be made of the best meat-producing animal they can find and immediately they can clone an entire herd of this 'super-animal'. Other constituencies are quick to point out, however, that such an application of cloning technology is not only unethical, but also impractical. The fundamental rule of selective breeding is to maintain a high enough level of genetic variation. When you narrow down the number of lines of animals for meat or milk production, you run risks of problems from inbreeding. While the clones themselves will not harm the gene pool, farmers hoping to breed the clones would be unsuccessful.

The most beneficial use for cloning that is currently being discussed and developed is in the realm of transgenic animals. Transgenics are special animals which have the DNA from other animals spliced with their own in order to produce desirable consequences. Some of these benefits include pigs with the human trait allowing them to produce human growth hormone, and cows and lambs that can produce medicines within their own milk that can be beneficial for sick newborns.

This latter application was what the scientists at Roslin were working on when they initially discovered their technique for cloning. Once a near perfect transgenic has been created, cloning technology could be helpful in reproducing this perfect animal. Such an undertaking must be done in limited numbers, because widespread cloning as shown will harm the overall gene pool. Another suggested use of cloning would be to replenish endangered species. This application is unlikely to solve the problem of endangered species as it would have the problem of too narrow a gene pool for the species to ever reproduce itself in the wild. However when it comes to the end of species on Earth for good, anything is worth a try.

'God alone is able to frame and make a living creature, fashion the parts, and mould and suit them to their uses.' (John Locke)

The concerns that have been raised by others about cloning are mainly theological. The Church of Scotland through their Science, Religion and Technology Project have released a document which outlines specifically the ethical concerns with animal cloning as well as the ways they believe it would be okay to use cloning technology. For the most part the Church does not have a problem with using these techniques for transgenic animals or most other medical applications for humans. The Church does warn that such technology should never be applied to animals that exist simply as food.

'In the limit, to manipulate animals to be born, grow and reach maturity for sale and slaughter at exactly the time we want them, to suit production schedules suggests one step too far in turning animals into mere commodities.' (Church of Scotland)

To date no country has any law making illegal the cloning of animals. Most feel that if it is okay to eat animals then it is certainly all right to clone them. In fact, the US patent office allows patents for such animals. They have issued patents for nine such animals since 1985. However, many politicians throughout the world have fears that the technology that has come about from animal cloning could be applied in the near future to human cloning. This has upset even the most stern advocates of animal cloning such as Dr Ian Wilmut of Roslin who has said that he hopes his work will never be applied to humans.

The foreseeable future of human cloning is easy to estimate. Research will continue in the field of medical applications of cloning, such as production of transgenic animals, and in research like that of the NIH. These groups have their feet solidly in the cloning market and are likely to maintain their status for at least the next 10-20 years. Other applications such as ABS's goal of superior livestock, are less likely to be thought of as necessary by the public. For that reason they will probably not be as successful in the improvement of cloning technology. As technology improves it will likely be impossible to limit any uses for animal cloning. Once techniques have been mastered that are ninety per cent efficient or better, the debate about cloning will be winding down. More than likely society will decide that benefits of cloning animals will outweigh any ethical implications. As improvements continue in the field of animal cloning it is impossible to tell how this technology could be applied 50 to 100 years from now. In this subject as with all new sciences the sky is the only limit on how far it can go.

• The above is an extract from the web site of the University of Virginia at www.cs.virginia.edu/

Commonly asked questions

By Lazaron BioTechnologies

Questions commonly asked by pet and other animal owners about cloning, DNA preservation and how it all works.

Q. Why should I save tissue/cells from my pet?
A. The scientists at Lazaron Bio-Technologies believe that cloning of pets and other animals will become routine within the next few years. In order to clone your pet, you must first preserve some genetic material (DNA) from the animal, which is located in every somatic cell of the body. Lazaron can preserve this DNA through cryopreservation technology.

Q. What kinds of animals are good candidates for cloning?
A. Although in theory, any mammal is just as good a candidate for cloning as another, the animal species with the most economic importance are most likely to be cloned first. Thus, domesticated animals (livestock, dogs, cats, horses and so forth) are the most likely to be good candidates.

Q. How much will it cost to clone my animal?
A. The answer is not clear. The very early clones were very expensive because they were a result of intensive research efforts. Just like a calculator or computer, the very first models were very expensive but as the technology developed, the costs dropped dramatically. We anticipate the same scenario with cloning. We believe that when cloning becomes routine, the cost of a cloned animal will not be much more than the cost of a quality pure breed.

Q. How do I submit a tissue sample to Lazaron BioTechnologies?
A. You can either contact Lazaron and we will send an Animal Tissue Retrieval Kit to your veterinarian or you can have your veterinarian contact us directly at + (1) 888 882 8918 or + (1) 225 334 6988.

Q. When should I have a tissue sample retrieved from my pet?
A. The tissue biopsy can be taken at any time during your pet's life, young or old. An optimal time to do the tissue biopsy may be when you have your pet vaccinated or take your pet to the veterinarian to be spayed or neutered. Although viable tissue can often be retrieved within 24 hours after death, this is not recommended.

Q. Will the age of my pet when the tissue sample is taken affect the viability of the cells?
A. No. There will be no difference in the cell viability regardless of the age of your pet.

Q. Can a tissue sample be taken if my pet is sick?
A. Yes. The DNA in your pet's cells will still be viable regardless of the health of the animal.

Q. Can a tissue sample be taken if my pet suddenly dies?
A. Yes. Your veterinarian can take a tissue biopsy within 24 hours after your pet passes away.

Q. Will my pet have to undergo general anaesthesia for the tissue biopsy procedure?
A. This will be determined by your veterinarian. In most cases, general anaesthesia will not be required. Depending on the disposition of the pet, the veterinarian may wish to sedate your animal and administer a local anaesthetic.

Q. How long will the tissue biopsy procedure take?
A. The tissue biopsy procedure should only take a few minutes.

Q. Will the tissue biopsy cause disfigurement of my pet?
A. No. The skin biopsy will be 6 millimetres in diameter, about the diameter of a pencil. The biopsy site (normally in the belly area) will be cleaned, shaved and disinfected by your veterinarian.

Q. Will my pet experience any long-term adverse effects from the biopsy procedure?
A. No. The small biopsy site should be healed within 1-2 weeks.

Q. Can I submit a tissue sample to Lazaron BioTechnologies without taking my pet to the veterinarian?
A. No. Lazaron BioTechnologies will only accept tissue samples that have been retrieved by a licensed veterinarian. Tissue samples must be submitted with a form signed by your veterinarian stating that the tissue originated from the pet identified on the form.

Q. How much does it cost to store my pet's sample?
A. The cost for Lazaron Bio-Technologies to process and cryo-preserve your pet's tissue sample is $500. There is a $10/month storage fee that will be automatically charged to your credit card.

Q. How much will the tissue biopsy procedure cost?
A. The price of the tissue biopsy procedure will be determined by your veterinarian. Generally the cost is between $50 and $150 depending on your animal's circumstances.

Q. What if I want to discontinue storing my pet's sample?
A. If you want to discontinue the storage, you simply notify Lazaron BioTechnologies using the Lazaron Specimen Withdrawal Authorisation Form.

Q. What if I want to get my pet's cells returned to me or shipped to another laboratory?
A. You may request to have Lazaron BioTechnologies ship your pet's cells in a liquid nitrogen shipper tank back to you or to another laboratory at any time. You also have the option to ship some of the cryovials of cells to another laboratory and keep some of the cryovials stored at Lazaron. The client will pay for all shipping of frozen cells from Lazaron to another location.

Q. How will the tissue sample be shipped to Lazaron Bio-Technologies?
A. A Styrofoam box with refrigerant packs and shipping medium will be shipped to your veterinarian with a prepaid, Next Day FedEx label. The tissue sample will be placed in a sterile test tube containing tissue shipping medium and antibiotics and returned by your veterinarian to Lazaron's laboratory on the campus of Louisiana State University.

Q. How will the tissue sample be processed by Lazaron Bio-Technologies?
A. When the tissue is received by Lazaron, the epidermis (outer layer) will be removed and the dermis (inner layer) will be processed by Lazaron's team of scientists. The dermis will be dissected into very fine pieces and placed in a tissue culture flask with a nutrient solution. Over a period of approximately 7-14 days, the cells of the dermis will begin to divide and grow on the bottom of the tissue culture flask. When there is sufficient quantity of cells, they will be frozen in a cryoprotectant solution and stored in a liquid nitrogen tank.

Q. What is the difference between tissue and cells?
A. Tissues are made up of individual cells. The sample taken from your pet will be a piece of skin tissue approximately 6 mm in diameter. Lazaron does not freeze this piece of tissue; we will dissect and culture the tissue in an incubator so that we can grow fresh cells and these cells will then be frozen.

Q. Why freeze cells instead of tissue?
A. The cryoprotectant (freezing solution) that is used will not penetrate the large tissue samples. It is much more efficient to cryo-preserve individual cells.

Q. What type of cells will be frozen?
A. Fibroblast cells (from the tissue sample) will be frozen. Fibroblast cells are the most common cell type that have been used to create the cloned animals that have been produced to date.

Q. How many cells will be frozen?
A. Approximately 1 million cells can be grown from the tissue samples received by Lazaron Bio-Technologies. These cells will be divided up into equal quantities and frozen in 3-5 cryovials.

Q. How do you count the cells?
A. The fibroblast cells are microscopic. A hemacytometer counting chamber on a microscope is used to count a fraction of the cells in the solution. Using a dilution formula, the total cell number can be calculated.

Q. How are the cells stored?
A. The cells, inside the cryovials, are stored on racks inside of a liquid nitrogen tank. Liquid nitrogen is -196° centigrade. This temperature is needed to ensure the cells are placed in suspended animation.

Q. Is this cryopreservation procedure considered experimental or a research technique?
A. No. Fibroblast cells have been successfully cryopreserved for decades. Scientists at Lazaron BioTechnologies have over 15 years' experience with freezing fibroblast cells from numerous tissues and many species of animals. Lazaron has further developed this technique and now offers this service to animal owners.

Q. What happens if the electric power fails in the Lazaron laboratory?
A. Power failures have no effect on the frozen cells. The cells are stored in liquid nitrogen tanks. These tanks do not rely on electricity to maintain their temperature. All of the incubators at Lazaron BioTechnologies operate on an uninterrupted power supply in the event of a power failure.

Q. Can my pet's cells get mixed up with another pet?
A. No. Lazaron BioTechnologies has a sophisticated labelling and record-keeping system, which assigns a unique identification number to your samples. Every dish, pipette, vial and tube is used only with individual samples to avoid sample-to-sample contamination.

Q. How does Lazaron Bio-Technologies protect my pet's sample?
A. Your pet's sample will be stored in a secured liquid nitrogen tank located in the Lazaron facility. This tank is also connected to a central alarm system that monitors the liquid nitrogen level.

Q. How long will my pet's cells remain viable in liquid nitrogen?
A. Cells such as spermatozoa, as well as embryos, have remained viable and produced offspring after being frozen and stored in liquid nitrogen for over 50 years. Cryobiologists believe that cells may retain their viability after storage in liquid nitrogen for thousands of years.

Q. Does Lazaron BioTechnologies provide the service of cloning?
A. No. Lazaron BioTechnologies does not provide the service of cloning. We believe that this service will be available in the near future. Scientists at Lazaron Bio-Technologies are involved in reproductive research (including cloning) at other institutions. Based on available data from other research organisations, Lazaron feels confident that cloning of pets will be a relatively routine procedure.

Q. Has anyone cloned a pet (dog or cat) yet?
A. No. However, Texas A&M University has received a $5 million grant from a private individual to develop the technology to clone a dog. This project was announced in September 1998 with a target success date of 2003. Texas A&M University's Reproductive Science Laboratory is well known in the scientific community and has a world-renowned research programme.

Q. Are cloned animals identical to the original animal?
A. Cloned animals are genetically identical to the original animal. However, the environment may influence factors such as behaviour and personality, much the same way it affects identical twins.

Q. Can I submit a tissue sample from outside of North America?
A. Yes, depending upon the country of origin and the donor animal circumstances. Call Lazaron at + (1) 888 882 8918 or + (1) 225 334 6988 to discuss your particular situation.

Q. Where can I get more information on the status of cloning?
A. Check out the information and links on Lazaron BioTechnologies' web site which can be found at www.lazaron.com/

• Lazaron BioTechnologies is a pioneer in providing genetic cryopreservation and related services to zoos, animal parks, breeders, trainers and other responsible animal owners. These services are provided in an effort to promote animal health and the protection of endangered, rare or valued animals through the application of new genetic technology. The above is an extract from their web site which can be found at www.lazaron.com/

Scientist aims to clone extinct Tasmanian tiger

By Aisling Irwin,
Science Correspondent

The extinct Tasmanian tiger could be brought back to life, say scientists, after the discovery of a perfectly preserved specimen in a museum.

The project, reminiscent of *Jurassic Park*, has been mooted by the director of the Australian Museum after he found a whole baby, born in 1866, preserved in a jar of alcohol. The animal became extinct 60 years ago but the specimen's DNA may be undamaged, said Prof Michael Archer. He also said that he has tracked down several other specimens in museums around the world. He said: 'There's a population waiting to be kick-started.'

The Tasmanian tiger, *Thylacinus cynocephalus*, is in fact a marsupial wolf with stripes along its back. The 5ft-long animal was widespread over Australasia until several thousand years ago, when it became confined to Tasmania. The last known example, Benjamin, died in a zoo.

Scientists have hoped that, should a full complement of perfect DNA be found, the genetic material could be inserted into the empty egg of the female of a similar species, which could then incubate it. The process would be a variation on the cloning of Dolly the sheep.

Prof Archer said that the sample was preserved in alcohol rather than the more destructive formalin, which gave him hope of perfect preservation. He said: 'We've discovered the miracle bottle in which this time capsule is just waiting to pop back into life.

'We've discovered the miracle bottle in which this time capsule is just waiting to pop back into life'

'At the rate at which this technology is increasing I wouldn't say that there's any reason why we shouldn't expect to be able to go into a pet shop and buy a pet thylacine and bring them home. We have cloning, we have DNA sequencing, we have the ability to read all the total information. If you like, the recipe for making a thylacine is there.'

But British geneticists dismissed the plan. Prof Martin Jones, archaeological scientist at Cambridge University, said: 'The DNA will be fragmented even though it is quite recently preserved.' Dr Richard Thomas, of the Natural History Museum in London, said that there was no living close relative of the Tasmanian tiger, and so successful incubation of an embryo was very unlikely. He said: 'There is a vast number of stages which we don't have the slightest idea how to do.'

ADDITIONAL RESOURCES

You might like to contact the following organisations for further information. Due to the increasing cost of postage, many organisations cannot respond to enquiries unless they receive a stamped, addressed envelope.

Animal Aid
The Old Chapel
Bradford Street
Tonbridge, TN9 1AW
Tel: 01732 364546
Fax: 01732 366533
E-mail: info@animalaid.org.uk
Web site: www.animalaid.org.uk
Animal Aid aims to expose and campaign peacefully against the abuse of animals in all its forms and to promote a cruelty-free lifestyle. Produces information including their quarterly magazine *Outrage*. To receive information on an issue or for a list of educational and information resources, please send a large sae to the address above.

BioIndustry Association
14-15 Belgrave Square
London, SW1X 8PS
Tel: 0171 565 7190
Fax: 0171 565 7191
E-mail: admin@bioindustry.org
Web site: www.bioindustry.org
Aims to foster greater public awareness and understanding of biotechnology and to encourage informed public debate about its development. If you have any questions or comments on the work of the BIA or their web site, contact them at the above address.

British Humanist Association
47 Theobald's Road
London, WC1X 8SP
Tel: 0171 430 0908
Fax: 0171 430 1271
Web site: www.humanism.org.uk
The British Humanist Association is the UK's leading organisation for people concerned with ethics and society, free from religious and supernatural dogma. It represents, supports and serves humanists in the United Kingdom and is a registered charity with more than fifty affiliated local groups. Publishes a wide range of free briefings including material on the issues of abortion and surrogacy.

Compassion in World Farming Trust (CIWF)
5a Charles Street
Petersfield
Hampshire, GU32 3EH
Tel: 01730 268070
Fax: 01730 260791
E-mail: ciwftrust@ciwf.co.uk
Web site: www.ciwf.co.uk
CIWF Trust is the educational wing of Britain's leading farm animal welfare organisation, Compassion in World Farming (CIWF), which has been campaigning for improvements in the welfare of farm animals for the last thirty years. Produces a Genetic Engineering & Farm Animals video. See the Educational Resources section on their web site, or contact them directly at the above address for details

Genetics Forum
94 White Lion Street
London, N1 9PF
Tel: 0171 837 9229
Fax: 0171 837 1141
E-mail: geneticsforum@gn.apc.org
Web site:
www.geneticsforum.org.uk
The Genetics Forum works to educate and inform interested parties about the implications of genetic engineering. Researches the uses and implications of genetic technologies on human health, the environment and animal welfare. It acts as an independently funded source of information for scientists, government, education and the general public. They publish *Splice* magazine. Ask for their publications list.

Human Genetics Advisory Commission (HGAC)
Office of Science and Technology
Albany House
94-98 Petty France
London, SW1H 9ST
Tel: 0171 271 2131
Fax: 0171 271 2028
Web site: www.dti.gov.uk/hgac

The HGAC was established in December 1996 to offer Government independent advice on issues arising from developments in human genetics. Amongst others, one of the topics that the Commission has addressed as a priority is cloning.

Society, Religion and Technology Project (SRT)
Church of Scotland
John Knox House
45 High Street
Edinburgh, EH1 1SR
Tel: 0131 556 2953
Fax: 0131 556 7478
E-mail: srtp@srtp.org.uk
Web site: www.srtp.org.uk
SRT is a project of the Church of Scotland set up in 1970 to examine ethical issues emerging from modern technology and to engage with key scientists and policy makers. It seeks to provide balanced and informed insights on major current issues. Its work includes genetic engineering, cloning, patenting, risk, environment, energy and 'God and Science' issues. It produces information sheets on a wide variety of issues which are available from the above address.

The Wellcome Trust
The Wellcome Building
183 Euston Road
London, NW1 2BE
Tel: 0171 611 8888
Fax: 0171 611 8545
E-mail: reception@wellcome.ac.uk
Web site: www.wellcome.ac.uk/
The Wellcome Trust runs many innovative activities for teachers and students, and offers a number of teaching resources. The teaching resources also help teachers stay up to date with the latest research in biomedical science, and detail practical activities that help to bring scientific subjects to life in the classroom.

INDEX

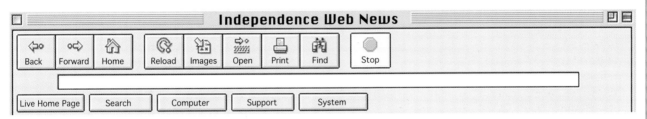

The Internet has been likened to shopping in a supermarket without aisles. The press of a button on a Web browser can bring up thousands of sites but working your way through them to find what you want can involve long and frustrating on-line searches.

And unfortunately many sites contain inaccurate, misleading or heavily biased information. Our researchers have therefore undertaken an extensive analysis to bring you a selection of quality Web site addresses.

* * * * *

Society, Religion and Technology Project
www.srtp.org.uk
This site aims to bring professional expertise by providing informed and penetrating comment for technologists, educators, media, the Church, the public – in fact anyone with an interest in how technology is affecting our lives, and the issues it raises. For their views on cloning, scroll down the home page to the Click on a Heading section. Click on Subject Heading. This takes you to their main index. Under the heading of Biotechnology is a link called Cloning. Click on this for a wide range of articles about human and animal cloning issues. Well worth a visit.

Human Cloning Foundation
www.humancloning.org
The Human Cloning Foundation (TM) promotes education about human cloning and other forms of biotechnology with an emphasis on the positive aspects of new technologies. A huge site with articles on a wide range of cloning issues including the benefits of human cloning and the reasons why we should clone human beings.

Roslin Institute
www2.ri.bbsrc.ac.uk
Click on Special Topic: Cloning. Then click on Background Notes on Nuclear Transfer and Cloning. This site includes all the background material produced by the Institute on cloning and nuclear transfer.

The Genetics Forum
www.geneticsforum.org.uk
The Genetics Forum is the only independent organisation in the UK concerned with the use of new genetic technologies and their public policy implications. It was founded in 1989 by a group of scientists, lawyers and advocates from the animal welfare, environmental and consumer movements concerned about the long-term impact of rapid developments in the genetic sciences. Since its inception, The Genetics Forum has been instrumental in raising the issues of genetic engineering to a prominent position on the public policy agenda.

MATCH
www.match.inweb.co.uk
MATCH is the first Scottish pressure group dedicated to campaigning against the cloning of human beings. Go to the MATCH Cloning Reference Page and click on the link Articles on Cloning to read the first of their articles on cloning.

International Cloning Society
www.suresite.com/wa/i/ics
The International Cloning Society (ICS) is essentially a body of people who have expressed an interest in being cloned at some point in the future. The Society is currently making arrangements to collect, store and preserve the cell/DNA specimens of those members who elect to make themselves available to be cloned, under specific and stated circumstances, in the future.

ACKNOWLEDGEMENTS

The publisher is grateful for permission to reproduce the following material.

While every care has been taken to trace and acknowledge copyright, the publisher tenders its apology for any accidental infringement or where copyright has proved untraceable. The publisher would be pleased to come to a suitable arrangement in any such case with the rightful owner.

Chapter One: An Overview

Cloning FAQ, © New Scientist 1999, *A history of cloning*, © Reprinted with permission from MSNBC Interactive News L.L.C., *Views*, © Kayotic Development, *Cloning techniques*, © Kayotic Development, *Cloning timeline*, © Kayotic Development.

Chapter Two: The Cloning Debate

How far should clone research go?, © Telegraph Group Limited, London 1999, *Breeding supergods*, © Lee Silver, 1998, *Human cloning*, © Christian Action Research and Education (CARE), *Cloning misconceptions*, © Kayotic Development, *Pros and cons of cloning*, © The Scotsman, June 1999, *The benefits of human cloning*, © Human Cloning Foundation, *Government rejects advice and blocks human cloning*, © Telegraph Group Limited, London 1999, *Forces that put cloning under a shadow*, © The Guardian, June 1999, *The clone guinea pigs?*, © The Daily Mail, February 1999, *Human cloning*, © NCGR, *Should we clone humans?*, © Donald M.

Bruce, 1998, *Anti-cloning research*, © Kayotic Development, *Cloning supporters*, Kayotic Development, *Send in the clones?*, © Brenton Priestley, *Ethical issues in cloning humans*, © Kate Cygnar from Minneapolis, USA, *Police warn of illegal cloning*, © The Independent, March 1999, *Cloning: is this the future for farm animals?*, © Compassion in World Farming (CIWF), *Cloning – how should society decide?*, © Donald M. Bruce, 1998, *Should we clone animals?*, © Donald M. Bruce, *$5m for lab to clone pet dog*, © The Guardian, August 1998, *First male mouse clone may save endangered species*, © Republished with permission from the Environment News Service (ENS), *Genetic flaws hit cloned animals*, © The Independent, April 1999, *Opinions on animal cloning*, © University of Virginia, *Commonly asked questions*, © 1999 Lazaron BioTechnologies LLC, *Scientist aims to clone extinct Tasmanian tiger*, © Telegraph Group Limited, London 1999.

Photographs and illustrations:

Pages 1, 12, 14, 18, 30, 32, 36: Pumpkin House, pages 2, 7, 10, 16, 21, 23, 26, 31, 35, 38: Simon Kneebone.

Craig Donnellan
Cambridge
January, 2000